BRITISH SCIENTIFIC LITERATURE IN THE SEVENTEENTH CENTURY

British Scientific Literature in the Seventeenth Century

Edited by

NORMAN DAVY

KENNIKAT PRESS
Port Washington, N. Y./London

BRITISH SCIENTIFIC LITERATURE
IN THE SEVENTEENTH CENTURY

First published in 1953
Reissued in 1970 by Kennikat Press
Library of Congress Catalog Card No: 71-105777
ISBN 0-8046-0947-0

Manufactured by Taylor Publishing Company Dallas, Texas

PREFACE

THE seventeenth century was pre-eminently that in which scientific research became experimental and therefore fruitful in new discoveries. The true method of investigation had been perceived and adopted at last. Splendid achievements followed one another thick and fast in Western Europe, and the philosophers of the British Isles, whom the public called virtuosi, contributed much. It was emphatically a lively, stirring period.

The purpose of this book is to recommend to those who are interested in the history of British science the works of some of those eminent men of the seventeenth century who described in English the results of their experimental researches and speculations about the nature of things. The first section is a historical introduction. Then follow some selected passages taken from original memoirs and books. No extracts from Latin works, even of such eminence as Newton's *Principia* or Gilbert's *De Magnete*, have been made, and this has caused the list of topics, as discussed in the historical introduction, to be different from that used in the selected passages.

It is hoped that the chosen extracts will not be found too technical for the general reader, and that the explanatory notes will be really helpful.

N. D.

University of Nottingham, 1953

CONTENTS

ACKNOWLEDGMENTS

PROFESSOR L. F. BATES has kindly lent me books concerning the early Fellows of the Royal Society. It also gives me great pleasure to thank Professor Vivian de Sola Pinto, for help given in the planning and compilation of this book; Miss D. Bexon, Miss T. M. E. Dawes, Dr W. W. Black, Mr W. R. Chalmers, Mr T. S. B. Hawthorne, Mr G. E. Trease, Mr H. A. Turner, and Mr F. Barnes, for technical assistance in preparing the explanatory notes; Mr G. E. Flack and the library staff of the University of Nottingham, for borrowing valuable books from external sources and making photostat copies; and Mr L. Heath. I would also thank Dr W. F. Pyle, of Trinity College, Dublin, for kindly supplying information about Irish virtuosi.

N. D.

INTRODUCTION

THAT the ancient Mesopotamians and Egyptians had more than a nodding acquaintance with what we now call pure and applied science is proved by the cuneiform inscriptions in the library of Assur-bani-pal (669–626 B.C.) at Nineveh and by the yet stronger evidence of the Egyptian Pyramids (*c.* 3000 B.C.).

Much of ancient Greek philosophy, which is the chief source of the intellectual tradition of Western Europe, was concerned with speculations on metaphysical and ethical subjects as well as logic and epistemology. Here the Greeks and their successors, the Romans, followed Socrates (*c.* 470–399 B.C.), who, according to his disciple Plato (*c.*427–*c.*347 B.C.), studied natural science in his youth, but found it unsatisfactory and devoted all his attention to what we should now call logic, ethics, and metaphysics. Aristotle (*c.* 384–322 B.C.), who was Plato's pupil, may be regarded as the chief representative of ancient science. He was the son of a physician, and an accurate observer of physical and biological data, though he apparently did little in the way of actual experiment or the testing of hearsay reports. There was, however, another tradition in Greek philosophy besides that of Socrates, Plato, and Aristotle. This was the tradition of the early Ionian and Sicilian thinkers (such as Thales and Empedocles), who were greatly interested in the nature and properties of the material world. Pythagoras the Samian (died *c.* 497 B.C.), living at Crotona in South Italy, studied the sounds emitted by vibrating strings. At a somewhat later date Democritus of Abdera (*c.* 460–357 B.C.) propounded the first atomic theory of matter. Archimedes of Syracuse (*c.* 287–212 B.C.) was certainly one of the great pioneers of experimental science. He is perhaps most famous for his principle, of buoyancy in connexion with bodies immersed in fluids, but he also invented a screw water-pump, established the theory of levers, and made important discoveries in the geometry of spheres, cylinders, parabolas, and spiral curves. Hero of Alexandria (*c.* 100 B.C.) wrote about machines such as pumps for raising water, about instruments such as theodolites and cyclometers, about military weapons such as catapults, and about mechanical toys. He even described a kind of steam-engine. Ptolemy (Claudius Ptolemeus), also of Alexandria (died *c.* A.D. 168), compiled all the then known facts of astronomy into a treatise now called the *Almagest*. In it he included a table of angles of incidence and refraction of rays of light.

Both experimental and theoretical science made little progress in the first six centuries after Christ. No discoverers of the first rank were born. Christianity spread, and when the Eastern Church became dominant in the empire whose metropolis was Constantinople, its leaders showed hostility to all forms of learning. At Alexandria the royal race of Ptolemys had died out. The Museum there, which was really a university endowed by them and supported by their patronage, ceased to turn out men and works of genius. There, too, the Church was bitterly opposed to science, and the woman mathematician Hypatia was murdered by a so-called Christian mob in A.D. 415. Finally the last traces of the Alexandrine University and Library were destroyed in the Mohammedan invasion of A.D. 641.

In the Dark Ages, following the closure of the philosophical schools of Athens in A.D. 529 and the murder of the last great scholar of antiquity, Boëthius, in A.D. 526, some experimental research was carried out by the Arabs of Bagdad, among whom the names of Geber, in chemistry, and Alhazen, in optics, were conspicuous.

In the Middle Ages, say from the ninth to the end of the fourteenth century, the finest intellects of Europe were under the spell of scholasticism. This was the combined study of theology and philosophy, particularly Aristotelian logic. With the exception of Roger Bacon, Jordan Nemorarius, and a few others, no one carried out new experiments. Charlemagne (A.D. 789) decreed the foundation of schools, in which instruction in the seven liberal arts of grammar, rhetoric, logic, arithmetic, music, geometry, and astronomy was given to the young. From the time of Gerbert, Pope Sylvester II, who died in A.D. 1003, the attention of the learned was directed towards the Greek masters, especially to Aristotle, and the subjects of the above curriculum were taught with reference to his authority and by the aid of oft-times corrupt versions of his works. Why appeal to experiment about a scientific topic when the *Physics* of Aristotle could be quoted? Experiments, therefore, were seldom carried out, and, in consequence, the Middle Ages had little scientific progress to record. However, towards the end of that period things began to happen. Universities were founded, gunpowder was discovered, printing was invented, Constantinople was captured by Mohammed the Second, and Columbus discovered America. Somebody somewhere constructed mechanical clocks, and somebody else glass spectacles.

By the year 1600 Europe was awakening, or awake, intellectually speaking. Superstitious veneration for the dicta of Aristotle was being replaced by a new passion for experiment. The world had been circumnavigated and proved to be round. Tycho Brahe had made catalogues of stars discovered by himself, and Gilbert was just publishing *De Magnete*. In the century between 1560 and 1660 there were living, besides great rulers like Elizabeth I and Henry IV of Navarre, and great writers like Shakespeare and Cervantes, such great pioneers of experimental science as Galileo and Harvey.

The lists of subjects taught in the universities of Europe in the sixteenth century seems to have included little besides divinity, Latin, Greek, Hebrew, philosophy, logic, rhetoric, civil law, canon law, medicine, pure mathematics, and astronomy. The last three were treated in the old-fashioned manner.

In the seventeenth century new topics appeared. The numerous voyages of navigators led to deepened interest in the world and to the study of geography, meteorology, and magnetism as well as the older astronomy and mathematics. To pure mathematics was added *mixed* or, as we now call it, *applied* mathematics. Astronomy led to advances in optics. The study of medicine led to physiology, anatomy, zoology, pharmacy, botany, chemistry, and physics. Mining led to mineralogy and geology. Astrology and alchemy began to die out. Dialling and horology were still fashionable.

In the tremendous progress in experimental science in the seventeenth century the natural philosophers of Italy, France, Germany, the Netherlands, Denmark, and Britain played the leading parts. Italy can show the great name of Galileo, who, by way of experiment, studied the laws of mechanics and applied the newly invented telescope to astronomical research; then followed Torricelli, Galileo's pupil, who invented the mercury barometer, and Malpighi, who used the then novel compound microscope to discover the capillaries conducting the blood stream. France boasts of Pascal, Gassendi, Pecquet, Mersenne, and Mariotte. Germany is proud of von Guericke, the burgomaster of Magdeburg, who invented air pumps and prepared vacua, and prouder still of Kepler, who designed the astronomical telescope, discovered the law of inverse squares in optics, and the laws of motion of the planets. The Netherlands claim Snell, the discoverer of the law of refraction of light, Stevin, the discoverer of the law of the triangle of forces, and Huyghens, the inventor of the pendulum clock and the propounder of the wave theory of light. Little Denmark can point to Römer, who first showed how to measure the velocity of light, and Bartholinus, who investigated the propagation of light in crystals.

A share, a great share, in the work was done in these islands. Though separated from the continent by the sea, the natural philosophers of Britain were in the forefront of the exponents of the New Philosophy, as it was called. The roll begins with the immortal names of Newton and Harvey, and it includes those of Boyle, Gilbert, Napier, Flamsteed, Halley, Ray, Hooke, and many others. Though Francis Bacon carried out no significant experimental researches himself, and though, at the time he wrote, the methods he advocated were already yielding fruit to others, his survey or prospectus of the New Philosophy aroused much interest throughout Europe and stimulated further effort.

The work of research was partly conducted by wealthy amateurs in their own houses. The Marquis of Worcester attempted to construct steam engines at Raglan Castle, Monmouthshire, and at Vauxhall; Robert Boyle investigated "chymicall" problems in his mansion at Stalbridge, Dorset, and studied the laws of gases contained in tubes erected in the staircase of his private lodgings at Oxford. Other distinguished gentlemen, amateurs in science, were John Evelyn, Kenelm Digby, Francis Willoughby, and William, Viscount Brouncker. The fiery Rupert, when Marston Moor and Naseby had become dim memories, found ease of mind in inventing explosive glasses or rediscovering the art of mezzotint. Even the indolent King, Charles II, found it a pleasant change from the labours of sauntering in the Green Park, or the scoldings of the Duchess of Cleveland, to descend into his 'elaboratory,' take part in a dissection, and 'hob-nob' with the virtuosi of the Royal Society.

A second class of investigators was composed of professors at the Universities. To obtain a chair at Oxford or Cambridge before 1600 a long course of study was necessary, even for men of genius. An undergraduate reading for a bachelor's degree in arts took grammar, logic, rhetoric, arithmetic, music, geometry, and astronomy in a course lasting four years. In the astronomy course there was no observatory and no telescope, nothing but the theoretical exercise of mastering Ptolemy. To proceed to the master's degree a three-years' course in metaphysics, moral philosophy, and natural philosophy was necessary. Natural philosophy consisted in reading the *Physics* of Aristotle. The doctorate, which required up to seven more years, could only be taken in the faculties of theology, law, or medicine. Thus William Gilbert, who was born in 1540, entered Cambridge in 1557, took the degrees of B.A. in 1561, M.A. in 1564, and M.D. in 1569. It appears that several of those who later on became experimental scientists had taken some kind of course in medicine, consisting of the reading of Hippocrates and Galen, along with a little botany and pharmacy. Such a person might have seen one or two dissections of a dog in his course, but not of a human body. Of the original founders of the Royal Society, Bathurst, Ent, Goddard, and Glisson were of this type.

If a man were so fortunate as to secure a lectureship or a chair he literally read his lectures in Latin. Few enjoyed the amenities of an 'elaboratory,' as it was then called, even at Oxford or Cambridge. Wilkins, the Warden of Wadham College, Oxford, exhibited instruments, including a waywiser,[1] a thermometer, and a monstrous magnet, in his lodgings at Wadham. Newton investigated the dispersion of light by glass prisms in his rooms at Trinity College, Cambridge. Along with the staffs of universities we may include the doctors of the Royal College of Physicians and the professors of Gresham College, both in London.

These two classes of individuals, then, contributed largely to experimental science. The seventeenth century is, however, remarkable for the foundation of learned societies, in various parts of Europe, in which more rapid progress could be made by the combined efforts of members and by mutual criticism and assistance than by isolated individuals. Italy has the honour of being the first country to advance in this way. The short-lived Academia Curiosorum Hominum, established at Naples

[1] A waywiser was an instrument for recording the distance travelled by a vehicle.

in the sixteenth century, was the first society in Europe to devote itself to experimental science, and was followed by the Accademia dei Lincei at Rome. This academy, so called from the lynxes on its 'device,' was founded in Rome in 1601 by Count Cesi to further the pursuit of natural science. Galileo and della Porta were members. It was disbanded in 1630 on the death of Cesi. The Societas Ereunetica was founded at Rostock, Germany, in 1622, the Collegium Naturæ Curiosorum at Schweinfurt in 1652, the very important Accademia del Cimento, or Academy of Experiment, at Florence in 1657, the Royal Society of London in 1662, and the Académie Royale des Sciences at Paris in 1666. In 1683 a number of gentlemen living in Ireland founded the Dublin Philosophical Society, with the same aims and rules as the Royal Society of London. Among the earliest members were Sir William Petty, Provost Narcissus Marsh of Trinity College, Dublin, William Molyneux, and St George Ashe. These various societies had different characteristics, but one common aim, the increase of knowledge by means of experiment. Some of them broke up after a short life, but the Royal Society of London and the Académie des Sciences of Paris are in full vigour at the present day.

The modes by which individuals and learned societies recorded the results of their experiments and preserved them for the benefit of posterity may be collected under three heads. The scientific literature of the period consists of books, the journals of learned societies, and the private letters of individuals. At the beginning of the century, and even after the Revolution, British scientific literature often disdained the use of English, and, as it now seems to us, pedantically preferred Latin, though the reason may have been its value as an international language. Gilbert's treatise, the first great British book on physics, was called *De Magnete, Magneticisque Corporibus, et de Magno Magnete Tellure Physiologia Nova.* Newton's masterpiece, one of the greatest achievements of the scientific intellect, was named *Philosophiæ Naturalis Principia Mathe-*

matica. Harvey's lectures, explaining his theory of the circulation of the blood, were published under the title *Exercitatio Anatomica de Motu Cordis et Sanguinis in Animalibus*. As the century went on Latin gradually fell into disfavour. This is illustrated by the letters written by the naturalist Ray and his learned correspondents, and collected and published by Derham in 1718. At the beginning of the book, very few letters are in English; at the end, very few are in Latin.

As for learned societies, the Royal Society published the researches of its Fellows as *Philosophical Transactions* in Latin or English. Sprat,[1] in his history of that society, tells us how it was determined that both discourses and memoirs should be made in the plainest, unadorned language, free from flowers of rhetoric.

Besides the literary works of our scientific countrymen in their native tongue, numerous translations of important modern continental authors were made, and some British authors, who published in Latin, had the honour of seeing their works quickly translated into English by their admirers.

We shall now proceed to survey briefly the works of British authors in various branches of experimental science in the seventeenth century. In the list are included zoology; physiology with anatomy; botany; chemistry; geology; astronomy; physics, comprising heat, light, sound, magnetism, electricity, and general properties of matter; and scientific instruments. The account will end with a brief reference to important translations.

ZOOLOGY

As we have seen, the seventeenth century was a period when all branches of science began to advance by the method of experimental study, which replaced the former unfruitful practice of simple quotation from classical authors. As a first example we will take zoology. In 1609 Charles Butler (died 1647), a parson, published in English a meritorious book

[1] See p. 31.

called *The Feminine Monarchie*, dealing with bees. This book was largely the result of close observation by Butler himself, and contained much new information mingled with references to Aristotle. A Latin compilation called *Insectorum sive Minimorum Animalium Theatrum*, edited by Thomas Moufet[1] (1553–1604), came out in 1634, but it had few new facts.

John Ray (1627–1705) and Francis Willoughby (1635–72), working together until Willoughby's death, brought out several Latin books on natural history, including *Ornithologiæ. Libri tres* (1676), *De Historia Piscium libri quatuor* (1686), *Historia Insectorum* (1710), *Synopsis Methodica Animalium Quadrupedum et Serpentini generis vulgarum* (1693), and *Synopsis Methodica Avium et Piscium* (1713). Ray and Willoughby were careful observers; they dissected animals and recorded anatomical details. They suggested new systems of classification. Ray's work in natural history was not so significant as in botany. The section on quadrupeds and reptiles was the most important of what he did. In it he adopted a classification very much like that of Aristotle, and, like the latter, included whales among the fishes. For the age, however, Ray and Willoughby's contribution to knowledge was considerable and paved the way for their successors. Two English treatises entitled *The Historie of Foure-footed Beastes* (1607) and *The Historie of Serpents. Or the Seconde Booke of Living Creatures* (1608) were published by Edward Topsell. It was a kind of adaptation of Gesner's *Historiæ Animalium*. Martin Lister (1638–1712) published the Latin *Historiæ Animalium Angliæ tres tractatus* (1678–81), *Historia sive Synopsis Methodica Conchyliorum* (1685–92), and several papers in the *Philosophical Transactions of the Royal Society*. He made some new observations, contributed some profitable criticisms, and assisted Ray and Willoughby in their work on fauna.

Robert Hooke's *Micrographia*, published in English in 1665, contained many new facts in natural history discovered by Hooke himself by the aid of the newly invented compound

[1] Thomas Moufet. Also called Moffett and Muffet.

microscope. The large-scale diagrams of fleas, lice, mites, spiders, moths, gnats, flies, and their various organs, drawn by the author, were a complete revelation to that generation, as we may see from the entry of Mr Samuel Pepys in his diary on January 21, 1665.

Thomas Willis (1621–75) wrote a Latin book called *De Anima Brutorum* (1672), in which, in addition to worthless speculations on the souls of animals, there was a valuable description of anatomical work on certain invertebrate animals such as the oyster, the crayfish, and the earthworm.

The two John Tradescants, father and son, set up and ran a museum of natural history in London, collecting rare creatures, shells, and minerals from abroad wherewith to furnish it. They also established a so-called physic garden for growing herbs useful in medicine. The younger John Tradescant published a catalogue of the contents of the museum under the title of *Museum Tradescantium* (1656). The efforts of such men stimulated public interest in several branches of science.

We may perhaps include in the list of British authors of scientific works John Johnstone (1603–75), the son of a Scottish father and a Polish mother. He was born at Sambter, in Posen, East Prussia, and after some schooling at Thorn in that province, studied at St Andrews, Cambridge, and Leyden. He published at Frankfort, in Latin, a "Historia Naturalis," consisting of *De Piscibus et Cetis* (1649), *De Avibus* (1650), *De Quadrupedibus* (1652), and *De Serpentibus et Draconibus* (1653). These four, illustrated with copper plates, formed a complete survey of the animal world as known at that time. Though much esteemed, they are said to have exhibited more learning than judgment.

In Scotland, Sir Robert Sibbald (1641–1722), an Edinburgh physician, quoted as an authority on antiquarian lore in the ever-memorable quarrel between Sir Arthur Wardour and Mr Jonathan Oldbuck in Sir Walter Scott's *Antiquary*, seems to have been one of the earliest writers on zoology. He published

Scotia Illustrata (1684), in which the fauna of Scotland are tabulated, and *Phalainologia Nova* (1692), giving an account of the Cetacea of British seas.

PHYSIOLOGY (WITH ANATOMY)

The reputation of British science in the field of research was greatly enhanced on account of the fundamental discovery, by experiment, of the circulation of the blood, in man and animals, by William Harvey (1578–1657). He first announced his novel doctrine in a course of lectures to students at the Royal College of Physicians, London, and later on (1628) published his lecture notes in Latin as *Exercitatio Anatomica de Motu Cordis et Sanguinis in Animalibus.* This treatise described perhaps the most important single discovery ever made in physiology by anyone in any country. Harvey also wrote another Latin work, *Exercitatio de Generatione Animalium* (1651), but as other researchers were able to use the compound microscope for their observations shortly afterwards, Harvey's work in this second topic was soon superseded.

John Mayow[1] (1645–79), besides being one of the founders of modern chemistry by his work on the gases of the atmosphere, also did some first-rate physiological research. Although he did not use the name 'oxygen,' which had not then been invented, he showed that there was a *spiritus igneo-ærius*, or 'fire-air,' a constituent of atmospheric air, which played an essential part in the act of respiration in animals as well as in ordinary combustion. He also showed that 'animal' heat is generated in the muscles. His work was published as *Tractatus duo, de Respiratione et de Rachitide* (Oxford, 1668), and as *Tractatus quinque Medico-Physici* (1674).

The chief merit of Thomas Sydenham (1624–89) was to insist on observation as the only source of knowledge of diseases. The signs and symptoms of each complaint must be

[1] See, however, p. 26

carefully studied and recorded. In this way he hoped to arrive at general laws concerning the course and best treatment for various complaints. The gout in his own system provided him with a field of observation close at hand. He wrote *Methodus Curandi Febres* (1666) and *Observationes Medicæ* (1676). Francis Glisson (1597–1677), a founder Fellow of the Royal Society, investigated experimentally and then wrote memoirs on the disease of rickets and the anatomy of the human liver, stomach, and intestines. Thomas Wharton (1614–73) studied glands and published *Adenographia* (1656), a valuable Latin treatise in which he compared the various glands of the human body. Thomas Willis (1621–75) investigated the anatomy of the human brain and the nervous system, recording his results in a book called *Cerebri Anatome* (1664). Richard Lower (1631–91) carried out, in 1669, the first direct transfusion of blood from one animal to another, and also extended Harvey's work on the physiology of the heart. His chief work is called *Tractatus de Corde* (1669). Sir Theodore Turquet de Mayerne (1573–1655), a fashionable physician patronized by royalty, left notes on medical cases of considerable interest. Clopton Havers (*c.* 1650–1700) seems to have been the first man to examine the anatomy of bones seriously. His treatise is called *Osteologia Nova or Some New Observations of the Bones and the parts belonging to them* (1691). Edward Tyson (1650–1708) studied and published illustrated accounts of the structure of the porpoise, the chimpanzee, the rattlesnake, and the opossum. The actual titles of the papers, in chronological order, are *Phocæna, or the Anatomy of a Porpess* (1680), *Anatomy of the Rattlesnake* (1683), *Carigueya seu Marsupiale Americanum, or Dissection of a Female Opossum* (1698), and *Orang Outang sive Homo Sylvestris or the Anatomy of a Pygmy compared with that of an Ape* (1699). Sir George Ent (1604–89), another physician, besides being a founder of the Royal Society, wrote the Latin *Apologia pro Circuitione Sanguinis* (1641), supporting Harvey, and other works furthering progress in medical science.

BOTANY

Progress in botany in Britain in the seventeenth century is largely connected with the names of Robert Morison (1620–83) and John Ray (1627–1705). Morison was the first Professor of Botany at Oxford while Ray was a free-lance. There was as yet no Chair of Botany at Cambridge. The work of these two men lay mainly in their attempts to systematize botany. Since the time of Theophrastus, who, in the third and fourth centuries before Christ, had divided plants into three groups—trees and shrubs, under-shrubs, and herbs—very little progress had been made towards a better classification. Morison published several Latin works, in none of which did he explain his method fully. However, an anonymous twelve-page (Latin) tract, published in 1720, described his system in detail, and showed that he based his classification of plants on the forms of their flowers and fruits. Ray lived long enough to publish, in Latin, three systems of classification, in 1669, 1682, and 1703, respectively, the last representing his final judgment. This last work was called *Methodus Plantarum Nova Emendata et Aucta*. In it, Ray retained the three groups of Theophrastus, but in considering the herbs he used the characters of flowers, fruits, leaves, and the number of seed leaves as the basis of classification. To two groups he gave the names of *monocotyledones* and *dicotyledones*, expressions which still survive though with a somewhat different meaning.

Robert Hooke (1635–1703), the physicist and inventor, was probably the first man in Britain to examine the structure of plants by the aid of the newly invented compound microscope. He discovered the cellular character of plant tissue and published diagrams in his book *Micrographia* (1665), written in English. He did not make a systematic study of botany. This task fell to Nehemiah Grew (1641–1712), who along with Marcello Malpighi (1628–94), a famous Italian, founded the study of plant anatomy. Using Hooke's method, Grew examined and described the different cells and fibres in various organs of a large number of plants, and studied the functions of those organs, thus making progress in plant physiology. He showed that flowering plants, like animals, have sexes, and realized the effect of habitat on plant life. He studied the structure of seeds. These and many other botanical researches were published in four treatises, all in English, of which the first was *The Anatomy of Vegetables Begun* (1672), the second *An Idea of a Phytological History Propounded* (1673), the third *The Comparative Anatomy of Trunks* (1675), and the last *The Anatomy of Plants* (1682). The *Anatomy of Plants* consisted of the combined second editions of the three earlier works, largely rewritten and enlarged.

About the year 1667 Sir Robert Sibbald (1641–1722) and Sir Andrew Balfour[1] (1630–94), two Edinburgh physicians, instituted a garden for the cultivation of plants giving drugs useful in medicine. This was situated near Holyrood Palace, and James Sutherland (*c.* 1639–1719) was placed in charge of it. Sutherland became the Professor of Botany at Edinburgh University, where a Chair was founded in 1673. He published *Hortus Medicus Edinburgensis* (1683), the first record of a collection of cultivated plants in Scotland.

CHEMISTRY

In the seventeenth century chemistry largely dissociated itself from alchemy and, so to speak, started in business on its own account as an independent branch of science. The leading chemist in Britain was undoubtedly Robert Boyle (1627–91). A Latin edition of his collected works was published at Geneva in 1677 and later, in 1744, a six-volume edition in English (second edition, 1772).

[1] Balfour was tutor to the poet John Wilmot, Earl of Rochester, and to the Earl of Ross. See *English Biography in the Seventeenth Century*, edited by V. de S. Pinto, in this series, pp. 99, 211.

Perhaps the most famous of the numerous memoirs included in these collections was *The Sceptical Chymist*, originally published in English in 1661. Among the new clear concepts or principles stated in this essay were the following: an important function of chemistry is to investigate the composition of substances in general, that is, to analyse them; only bodies which cannot be decomposed further are to be regarded as elements; many new elements remain to be found; the affinity of certain elements for others and the approach of fine particles of one to the other leads to the formation of compounds; the properties of a compound are different from those of the constituent elements. Macaulay's famous expression "Every schoolboy knows so-and-so" is perhaps most nearly true in the case of Boyle's law relating to perfect gases. Boyle himself wrote about "the hypothesis that supposes the pressures and expansions to be in reciprocal proportions" in a memoir, *New Experiments Physico-Mechanical touching the Spring of the Air*, in 1659. His word 'expansion' means the same as our word 'volume.' It has been stated that the French scientist Edme Mariotte rediscovered the same law in 1676. Boyle tried to find out the true nature of the atmosphere but failed.

The short-lived John Mayow (1645–79), also notable for his physiological work on respiration, made some progress towards the solution of that problem. His ideas and results were published in Latin in 1674, under the title *Tractatus quinque Medico-Physici*. In this work Mayow brought evidence to show that atmospheric air consisted of at least two kinds of gases. One of these, which he called *spiritus igneo-ærius*, or 'fire-air,' supported life and combustion. It was present in saltpetre. The other constituent was lighter than 'fire-air' and did not support life and combustion. Thus Mayow had almost discovered oxygen and nitrogen, and had studied some of their properties. He also prepared a new gas, now called nitric oxide, by the action of nitric acid on iron. This gas, he found, when introduced into ordinary air, confined over water, diminished its volume. All this shows what a first-rate experimenter and inferential reasoner Mayow was. It should, however, be stated that two competent modern authorities, Professor T. S. Patterson and Dr J. F. Fulton, have published articles in the journal *Isis*, asserting that Mayow's work was not truly original but derived from that of Boyle.

Robert Hooke (1635–1703), in Observation XVI, "Of Charcoal, or burnt Vegetables" in his book *Micrographia* (1665), pointed out that in the process of combustion there was a necessary ingredient, namely "a substance inherent, and mixt with the Air, that is like, if not the very same, with that which is fixt in salt-peter." Fresh supplies of air increased the rate and violence of the action. Hooke's conception of the phenomena occurring inside flames was very much like the modern one. One Christopher Baldwin (1632–82) noticed the phosphorescent property of glowing in the dark possessed by anhydrous calcium nitrate.

GEOLOGY

In the Stuart period British geology did not make much progress. Thomas Burnet (*c.* 1635–1715) was one of the exponents of a religious theory of the formation of the earth. He published in Latin a book called *Telluris Theoria Sacra* in 1681, giving a most fanciful account of the origin of the globe in a chaos of four components, earth, water, oil, and air, and describing its subsequent evolution to the present state. At one stage the wickedness of man caused the shell of the egg-like earth to burst so that water emerged and produced Noah's Flood. In 1684 a translation called *The Theory of the Earth: containing an Account of the Original of the Earth, and of all the General Changes which it hath already undergone or is to undergo* was published. John Woodward (1665–1728) wrote in English. His principal work was the *Natural History of the Earth and Terrestrial Bodies* (1695), in which he made some speculations on the probable changes which had occurred in the earth's surface formation from the period before the Deluge to the time of writing. There was something of permanent value in Woodward's treatise, that is, his description of many fossils, and the then novel suggestion that fossils represented past animals and plants. He insisted that certain stone objects then called *Ceraunii*, were man-made tools and implements and not natural bodies. William Whiston (1667–1752), in the *New Theory of the Earth* (1696), let his imagination run riot in wild speculations about the origin of the earth. Much more significant were the views of the versatile Robert Hooke (1635–1703), which were published as "A Discourse on Earthquakes" in a volume called *Posthumous Works* (1703). Hooke displayed a critical modern attitude towards geology, as towards all other branches of science. He regarded fossils as records of former life, either petrified organisms or impressions left by them, and, after examining many specimens most carefully, made very fine drawings of them. He pointed out the importance of earthquakes as disturbers of geological strata, and regarded them as responsible for lifting fossils to great heights above sea-level. Hooke predicted the use of fossils as a means of arranging strata in a true chronological order. Martin Lister (*c.* 1638–1712) would not accept the view that fossil shells had ever belonged to living animals, and John Ray (1627–1705) could not make up his mind to accept any definite conclusion about their origin. In *Three Physico-Theological Discourses* (1693) he emphasized the significance of running water as a producer of surface erosion. George Sinclair (died 1696), a Scot, published in 1672 a work called *The Hydrostaticks*. It included a section called "A Short History of Coal, and of all the Common and Proper Accidents thereof; a Subject never Treated of Before." This was a description of the Midlothian coalfield and represented the beginning of the systematic study of Scottish carboniferous stratigraphy. The geometrical treatment of the strata was most able, but fossils were never mentioned. The same

Martin Lister mentioned above also suggested the construction of geological maps to exhibit types of rocks and soils to be found throughout the country, but he does not appear to have drawn one.

ASTRONOMY

In theoretical astronomy and the subject of the general properties of matter, the discovery of the universal law of gravitation by Isaac Newton (1642–1727) must be reckoned one of the most important scientific events not only in the seventeenth century, but in the whole history of human thought. According to that law, every particle of matter in the universe attracts every other particle with a force proportional to the masses of those particles and to the inverse square of the distance apart. Kepler's three laws governing the motion of the planets were seen to be the logical consequence of Newton's single law, which also provided an 'explanation' of the motion of our moon, of the satellites of other planets, and of certain comets. Halley, Hooke, and Wren had all made the happy guess that the law governing the motion of the planets was that of the inverse square of the distance, but none of them could calculate the shape of the orbit of a planet obeying that law. Newton had solved the problem in 1666, showing that such an orbit should be an ellipse, but when he tried out his theory with the faulty data then available it did not seem to be satisfactory and he remained silent. In 1682 he procured some improved data on the dimensions of the earth, newly measured by the Frenchman Picard, and using these found that his law would account for planetary motion perfectly. Newton submitted his work to the Royal Society on April 28, 1686, in a paper forming what later appeared as the first book of the famous Latin treatise *Philosophiæ Naturalis Principia Mathematica*, published in 1687.

John Flamsteed (1646–1719) was in the main a practical

astronomer who became the first Astronomer Royal of England, and who catalogued accurately the positions of about 3000 stars. His chief work, the *Historia Cœlestis Britannicæ*, was published posthumously in 1725. Edmund Halley (1656–1742) was both a theoretical and practical astronomer as well as a physicist of high rank. He discovered what are called the proper motions of the stars, the secular or long-period acceleration of the moon's mean motion, and the periodic nature of the motion of comets. As is well known, his name is attached to that comet which appeared in 1682 and whose return, according to his calculations, was to be expected in 1758, and later, with equal time intervals. It did appear on Christmas Day 1758. The interval of nearly seventy-seven years was his calculated value of the period of the comet, and by subtracting one, two, three, and greater integral multiples of this from 1682, he showed that the same comet had already appeared many times in the world's history. After a voyage to St Helena, in which he observed about 300 stars visible in the southern hemisphere only, he published a *Catalogus Stellarum Australium* (1679). If Jeremiah Horrocks (1619–41), a Lancashire curate, had lived longer he might have discovered much, for by his own calculations and corrections to Kepler's tables, he predicted that a transit or passage of the planet Venus across the sun's disc would occur on Sunday, November 24, 1639. His prediction was fulfilled exactly and the transit observed by Horrocks and his friend Crabtree. James Gregory (1638–75), in his *Optica Promota* (1663), indicated a method of measuring what is called the sun's or solar parallax. This is the ratio of the earth's radius, measured at the equator, to the mean radius of the earth's orbit. The method involved observation of transits of the planets Venus or Mercury across the sun's disc. Gregory also suggested a method of measuring the parallax of the fixed stars in a work called *Geometriæ Pars Universalis*, published at Padua in 1668.

PHYSICS

Heat. In the seventeenth century the subject of heat was in its infancy, and, further, did not grow very fast. One of the first, if not the very first, thermometers was constructed in Italy by Galileo in the earliest years of the century, and a Dutchman, Drebbel, made a thermometer consisting of a glass bulb containing spirits of wine, with an emerging straight tube, about the year 1620. The civil war hindered progress in these islands, but from 1660 systematic attention was given by members of the Royal Society to various topics which can be classed under the title of heat. Sprat's *History of the Royal Society* (1667) tells us of the invention of

> several kinds of Thermometers for discovering the heat and cold of the Air, or any other Liquors: A Thermometer for examining all the degrees of heat in Flames and Fires, made of several substances; as also the degrees of heat requisite to melt Solder, Lead, Tin, Silver, Brass, Iron, Copper, Gold.

Sprat also states that the Royal Society has considered methods

> of making a Standard of Cold by freezing distill'd Water, of producing Cold by the Dissolution of several Salts. It has carried out experiments of the Expansion and Condensation of Water by Heat and Cold, of converting Water into a vaporous Air, lasting some time in that Form [and experiments] to determine the possible Bounds of Expansion and Condensation of the Air by Heat and Cold, by exhausting and compressing.

He says that Sir Christopher Wren has invented a self-registering thermometer, and a circular thermometer in which a certain error is eliminated. A perusal of Robert Boyle's *Works*, first published collectively in English in 1744 but containing memoirs originally published in the years 1660–70, shows that Boyle himself was one of the chief workers who had carried out researches of the kind just mentioned. According to Sprat, the Royal Society demanded from its fellows in their spoken and written reports an economy in words and plainness in style not previously the fashion.[1] It had decided

> to reject all the Amplifications, Digressions, and Swellings of Style; to return back to the primitive Purity and Shortness, when Men deliver'd so many Things, almost in an equal Number of Words. They have exacted from all their Members, a close, naked, natural way of Speaking; positive expressions, clear senses; a native Easiness; bringing all Things as near the mathematical Plainness as they can; and preferring the Language of Artizans, Countrymen and Merchants, before that of Wits or Scholars.

Boyle obeyed some but not all of these instructions. His memoirs use plain language, free from any turgidity of style. The reader understands what the author means, and plates of diagrams help the technical discussions. There are, however, too many words in each paper, partly because of the author's anxiety to quote all the old information he can get, relevant to the matter in hand. In dealing with the "History of Cold," for example, he frequently quotes the experiences of the Dutch sailors stranded in Nova Zembla as described by Gerat de Veer, which Addison perhaps had in mind in composing No. 254 of the *Tatler*, dealing with the Frozen Words, unless he borrowed the idea from Rabelais.

In Boyle's *Works* we note that φαινόμενα is still a Greek word, that æquilibrium and adhæsion still contain a diphthong, the handle of an air-pump is called a manubrium, the piston a sucker, mercury at rest is said to be restagnant, adulterated materials are called sophisticated, a favourite vessel to hold gases is a limber lamb's bladder, glass bulbs are glass eggs or bubbles, oscillatory motion is called brandishing motion, experiments are not repeated but reiterated, difficulties are not explained but explicated, the level of a mercury column is said not to be considerably higher but a pretty deal higher, and a small quantity of air is called a parcel. For our word quickly

[1] See above, p. 18.

Boyle uses nimbly; for opaque, opacous; for trigger, tricker; for retort, bolt-head, helmet, or alembic; for short notes, advertisements; for intimately mixed, exquisitely incorporated; for a set of bodies, a congeries or heap; for suitable, commodious; for exhaustion of the air from a vessel, exsuction; and for the escape of the air, the recess. These few examples must suffice to remind us of the differences between the English of the middle of the seventeenth and the middle of the twentieth centuries.

Robert Hooke, particularly in the English work *Micrographia* (1665), showed marked power in the realm of heat, not only by using the freezing-point of distilled water as a standard lower fixed point of his alcohol thermometers, and by his accurate method of calibrating those instruments, but in his views about heat in general. Like Francis Bacon, he had the modern conception that the heat content of a body was "nothing else but a very brisk and vehement agitation of the parts of the body." In another place he said that "heat is a property of a body arising from the motion or agitation of its parts." Hooke also realized that the property of expansion and contraction with temperature changes was a general property of matter.

In the branch of applied heat Edward Somerset, second Marquis of Worcester, suggested a kind of steam-engine for raising water from a low to a high level ("driving up water by fire"). It is Invention No. 68 in his book *Century of the Names and Scantlings of such Inventions as at present I can call to mind to have tried and perfected*, published in 1663. It is not clearly settled whether such an engine was ever actually constructed. Thomas Savery (*c.* 1650–1715) actually made a steam-engine for pumping water out of mines, taking out a patent for it in July 1698 and describing it in a book, *The Miner's Friend*, published in 1702. Denis Papin (1647–*c.* 1712), a Frenchman who lived for some time in England, devised an experimental steam-engine for raising water on a laboratory scale only. His chief work, however, was the invention of a so-called "digester," described in a paper in the *Philosophical Transactions* (1680), "A New Digester or Engine for softening Bones."

LIGHT. In the century under consideration, progress in optics was much greater than in heat. Stimulated probably by the successful work of Galileo with refracting telescopes, and by Snell's discovery of the true law of refraction (1621), the leading experimental and theoretical scientists of Western Europe devoted much attention to this subject. In Britain the Scot James Gregory (1638–75) invented a reflecting telescope (1661), which he described in Latin in a book called *Optica Promota* (1663). This instrument consisted of a concave spherical mirror facing the object to be viewed. This mirror received and reflected the incident rays to a second small concave mirror mounted on the axis of the first, in front of and facing it. The final image was observed by means of an eyepiece piercing the centre of the larger mirror and directed towards the smaller one.

In England, Robert Hooke (1635–1703) made fundamental advances in that branch of optics called interference, publishing his results in English in the book *Micrographia* (1665). There he described experiments on the colours of thin films, plates, and soap bubbles, and gave a partly correct theory of them. He observed the phenomenon called Newton's Rings before Newton did. Speculating on the nature of light itself, he assumed the existence of a universal all-pervading ether, and thought that the passage of a ray of light through this ether consisted in the setting-up of rapid oscillatory motion of the particles. He even supposed the vibrations to be transverse and only went astray in supposing them to become oblique in other media. His theory of the production of coloured rays from white light is not clear. Hooke was the first to work out what is called the resolving power, or power of discernment, of an optical instrument, which he did for the human eye, getting a result in fair agreement with modern values. The discovery by

Francesco Grimaldi, an Italian Jesuit, of the phenomenon of diffraction, or bending of light round corners, had been published in 1665, and Hooke, seven years later, rediscovered it independently, making several experiments thereon and forming correct views. Hooke published his work on this subject in the *Philosophical Transactions of the Royal Society* in 1672 and 1675.

Isaac Newton (1642–1727), whose labours are described in his English work *Opticks* (1704), made a discovery of first-rate importance—namely, that sunlight was a mixture of rays of many colours. This he proved in a series of experiments beginning with a repetition of Descartes' experiment of passing sunlight through a glass prism, thereby causing its dispersion or separation into a coloured spectrum. The work was done in 1666, and formed the subject of Newton's first memoir, published in 1672 in English, in the *Philosophical Transactions of the Royal Society*. It was called "A New Theory about Light and Colour." In 1668 Newton invented a reflecting telescope, somewhat differing from Gregory's. The first image was formed by a concave mirror as in his instrument, but the second image was formed by a small plane mirror set at 45 degrees to the axis of the concave mirror, and it was viewed by an eyepiece inserted laterally into the tube. The published description was called "An Accompt of a New Catadioptrical Telescope" (1672). In a paper published in 1675 Newton supported an ether theory of light. According to him, however, it was not vibrations of the ether particles which constituted rays of light, but "small bodies emitted every way from shining substances" (Query 29 of the *Opticks*). By the scientific world this corpuscular theory was long supposed to be completely unsound, and superseded by the wave theories of Huyghens and others, but by the advent of the quantum theory, it recovered much of its prestige. In the *Opticks* (1704) Newton considered the colours of thin films, and to explain them invented a theory of "fits of easy reflexion and transmission," a theory in which there is some confusion of thought. He, as it were, was determined to avoid the use of the concepts of the wave theory, though if for Newton's phrase "length of fit" one substituted the expression 'wavelength,' this theory would almost become that of modern textbooks. In the second edition of the *Opticks* (1717) Newton expressed views in which there is a clear recognition of the polarity of the two rays of light, obtained by double refraction in Iceland Spar.

Robert Boyle (1627–91) wrote on optics in "The Experimental History of Colours" (*Works*, Vol. I). This paper is a collection of experimental observations either of Boyle himself or others, but whereas Newton investigated certain topics most thoroughly, devising further quantitative experiments to test his views, Boyle seldom went further than describing simple experiments and making a few qualitative deductions from them. Boyle noted, for example, that black surfaces absorbed heat better than white ones, that persons about to be attacked by the plague saw surrounding objects clothed in glorious colours, that on skimming molten lead the clean surface soon began to display successive bright colours, that a certain blind Dutchman could distinguish coloured bodies by touch, that a 'parcel' of new snow could not be distinguished in a dark room, and so on.

Edmund Halley (1656–1742), the astronomer, is said to have been the first person to publish the now well-known equation connecting the focal length of a thin convex lens with the object and image distances. It appeared in the appendix to a treatise called *Dioptrica Nova* (1692), by William Molyneux (1656–98), a founder member of the Dublin Philosophical Society. David Gregory (1661–1708), nephew of the Scottish James Gregory mentioned above, published in 1695 a Latin work called *Catoptricæ et Dioptricæ Elementa*, in almost the last line of which is to be found the basic notion of achromatic combinations of lenses, a concept which had escaped even the genius of Newton.

SOUND. A remarkable advance in the branch of theoretical acoustics was made by Isaac Newton (1642–1727). He arrived at clear ideas as to the phenomena involved in the propagation of sound through a uniform medium. They are expounded in Latin in the second volume of the *Principia* (1687). Newton's conception of longitudinal waves of sound, consisting of trains of alternate condensations and rarefactions, is still unchallenged. In one point, however, Newton did not quite reach the whole truth. He envisaged vibrations of the medium as taking place under 'isothermal,' that is, constant temperature, conditions, and his theory, when applied to atmospheric air, gave a value of the velocity of sound which disagreed with experimental values. It was not until 1817 that Laplace showed that the vibrations occurring in sound waves were so rapid that they must be regarded as 'adiabatic'—that is, occurring without any transfer of heat to or from the portion of medium traversed by the sound. Newton's expression for the velocity of sound in air, thus corrected by Laplace, gave results in good agreement with experiment.

Robert Hooke (1635–1703), with his usual skill, studied the sounds emitted by toothed wheels and glass bells, and measured the velocity of sound in iron wires. His friend, Mr Samuel Pepys, notes in his diary on August 8, 1666, the excellent quality of Mr Hooke's discourse on the musical sounds made by vibrating strings and insects' wings. Two Oxford Fellows, William Noble and Thomas Pigot, probably working under the direction of the mathematician John Wallis (1616–1703), made some interesting observations on the transverse vibrations of stretched strings, introducing the use of paper riders in such studies, but they did not hit upon the true law of such vibrations. Their work, done in 1674, was described in English in the *Philosophical Transactions* for 1677, under the title "Of the Trembling of Consonant Strings," by Dr John Wallis. Sprat tells us that the Fellows of the Royal Society had invented before 1667 several kinds of "Otocousticons or

Instruments to improve the sense of hearing" and had studied "Echoes and reflected Sounds, musical Sounds and Harmonies." He was probably referring to work done by Hooke. In 1670 Sir Samuel Morland (1625–95) invented a speaking-trumpet, or megaphone, which was called a "stentorophonical tube."

MAGNETISM. The year 1600 saw the publication in Latin of an epoch-making book called *De Magnete Magneticisque Corporibus et de Magno Magnete Tellure Physiologia Nova*, by William Gilbert (1540–1603), physician to Elizabeth I. By the work of Gilbert, magnetism became a true branch of science. The leading novelty in his conceptions was that the whole earth was one huge loadstone or magnet, with what we now call a magnetic field[1] around it. Evidence supporting this notion was brought forward. An iron sphere, called a terella, was prepared and magnetized parallel to a diameter, and thus became a model of the earth. Small magnets, suspended by thin fibres near the terella, set in definite directions like freely suspended magnetized needles do near the earth's surface. Iron rods placed near the terella became magnetized. The terella possessed two poles like the earth. Thus Gilbert justified his hypothesis and earned the warm commendation of Galileo for the step forward. The great poet Dryden expressed the general view in his Epistle to Dr Walter Charleton, physician to Charles II, "Gilbert shall live till loadstones cease to draw." The high place occupied by Gilbert in the history of science is due to the fact that he carried out experiments for himself, in the manner laid down by both Roger and Francis Bacon, and that he was both willing and able to make true deductions from the result of those experiments.

Edmund Gunter (1581–1626) continued Gilbert's work by making regular observations of the angle between the direction in which the axis of a horizontal compass needle sets and a true

[1] Newton and others called a 'field' of this, or of the electrical kind, an *effluvium*.

geographic horizontal north and south line, and by showing that this angle was gradually changing through the years. Edmund Halley (1656–1742) saw that the study of terrestrial magnetism would throw light on the structure of the earth, and he devised a theory establishing a connexion between the two. He made journeys to St Helena and other places in the southern hemisphere in 1677 and 1700 in order to record magnetic data. He showed that the luminous displays of the aurora borealis (or australis) were closely connected with terrestrial magnetism. Robert Boyle (*Works*, Vol. I) proved by experiment that magnetic forces acted as well across a vacuum as through air. Although his expression is "that in a vacuum the loadstone . . . attracted or repelled . . . according to the laws magnetical," it does not seem that he really knew the law of inverse squares. Isaac Newton mentioned in the *Opticks* that

> the Magnet acts upon Iron through all dense Bodies not magnetick nor red hot, without any diminution of its Virtue, as for instance, through Gold, Silver, Lead, Glass, Water.

We may also mention here two other works on magnetism. In 1613 Dr Mark Ridley (1540–1624), who had lived in Muscovy as physician to the false Czar, Boris Godounoff (Czar from *c.* 1598 to 1605), published *A Short Treatise of Magneticall Bodies and Motions.* In it he discussed such topics as magnetic declination and dip, and methods of measuring them.

Shortly afterwards, in 1616, William Barlowe (*c.* 1540–1625) published *Magneticall Advertisements: or Divers Pertinent observations, and approved experiments concerning the nature and properties of the Load-stone: Very pleasant for knowledge, and most needfull for practise, of travelling, or framing of Instruments fit for Travellers, both by Sea and Land.* Barlowe is said to have discovered the difference between the magnetic properties of iron and steel, and methods of making magnets by touch. In his book Barlowe accused Ridley of having stolen his manuscript and of incorporating the information in his (Ridley's) book. He boasted of certain priority and equal merit to Gilbert.

ELECTRICITY. The same William Gilbert was the father of electrical as well as of magnetic science. In the same work *De Magnete* (1600) he reported experiments in what we now call electrostatics. Gilbert observed the attractive forces exerted on light bodies by amber, glass, sulphur, sealing wax, jet, alum, and precious stones when these substances had been excited by rubbing. He showed that the presence of moisture reduced the vigour of the forces manifested, and that an electrified body was discharged if a live coal was brought near. In these experiments he employed a kind of electroscope which he had invented. Gilbert distinguished between the forces of electrified and those of magnetized bodies.

After the work of Gilbert no further progress seems to have been made in this subject in these islands in the seventeenth century.

GENERAL PROPERTIES OF MATTER. After Newton's law of gravitation and Boyle's law of gases, discussed elsewhere in this account, perhaps the most important general law concerning the properties of matter discovered in Britain in the seventeenth century was Hooke's law governing the extension of elastic cords and spiral springs. An account of this was published in 1678 in English in a book *Lectures de Potentia Restituva; or of Spring*, though the discovery had been announced in 1676 in the form of the anagram *ceiiinosssttuu* of the Latin sentence *ut tensio sic vis*, counting the letter *v* as *u*. Thus the law stated that the amount of extension was proportional to the stretching force. Hooke's account of it was illustrated by clear figures of bodies under tension, a straight wire, a long helical spring, a watch spring, and a cantilever rod. Boyle's law for gases was quoted as another example, and, indeed, Professor Andrade thinks that Hooke was the actual discoverer of 'Boyle's' law.

A little progress was made in the branch of capillarity

Boyle (*Works*, Vol. I, p. 80) studied the rise of water in fine tubes. Hooke, in a fifty-page book called *An Attempt for the Explication of the Phenomena Observable in an Experiment Published by the Honorable Robert Boyle* (1661), published a theory which, though wrong, had the great merit of regarding the ascent of liquids in capillary tubes, or along the edges of partly immersed solids, or up the sides of a vessel, and the spherical form of drops, as examples of one general property. He also said that the same property of liquids was responsible for the rise of sap in trees and of oil in the wicks of lamps.

SCIENTIFIC INSTRUMENTS

The seventeenth century was characterized by a flood of inventions of instruments useful in scientific work. The most prominent of British inventors was Robert Hooke (1635–1703). The list of his achievements is immense. We will mention a few. To further the science of meteorology Hooke invented the wheel mercury barometer, the wind gauge, a self-measuring rain gauge, and a hygrometer. In other branches he invented the anchor escapement of clocks, the balance spring of watches, telescopic sights for a mural quadrant, a clock-driven telescope, a universal mechanical joint, and a system of optical telegraphy. These were described mainly in papers in the *Philosophical Transactions*. William Oughtred (1575–1660) and Edmund Gunter (1581–1626) both made contributions to the invention of the slide rule. The former is said to have first used the × sign used in arithmetic to indicate multiplication, and Gunter invented the surveyor's chain. His principal work was *The Description and Use of the Sector, the Crosse-Staffe and other Instruments* (1623). Sir Samuel Morland (1625–95), in an English book *The Description and Use of Two Arithmetick Instruments* (1673), described his machines for adding and subtracting pounds, shillings, and pence. In 1667 Richard Townley and Robert Hooke described in *Philosophical Trans-*

actions a micrometer screw, invented by "a Mr William Gascoigne (*c.* 1612–44), unfortunately slain in his late majesty's service." Townley also tells us that this brave Cavalier had a treatise on optics ready for the press when he was killed at Marston Moor on July 2, 1644.

TRANSLATIONS

Among the translations into English made during this period we may distinguish two groups: those of older classical authors, mostly Latin, Greek, and Arabic; and those of foreign and domestic authors describing recent researches in Latin, Italian, or French.

In the first set come the works of Philemon Holland (1552–1637), the "translator general in his age." In 1601 he published a complete version of *The History of the World, commonly called The Natural Historie of Gaius Plinius Secundus* (second edition, 1634), a full account of ancient science up to the time of Pliny (*c.* A.D. 23–79). In 1678 Richard Russell "faithfully Englished" the works of the eighth-century Moorish mathematician Abu Musa Gabir ben Hayan (*c.* A.D. 722–813), better and more simply known as Geber.

The second type is represented by a translation of the *Dialogo sopra i Due Massimi Sistemi del Mondo, Tolemaico e Copernicano*, of the famous Italian Galileo Galilei, by Thomas Salusbury, in 1661. This was the work which had brought Galileo into disfavour with the Inquisition after its publication in 1632, the work showing Galileo's views on the nature of the solar system. In 1662 John Chandler published *Oriatricke or Physick Refined*, a translation of the Dutch chemist Van Helmont's *Ortus Medicinæ*, first published in 1648 in Latin. This was the work in which Van Helmont had expounded his famous concept of invisible forms of matter, for which he had specifically invented the new word 'gas.' John French, in 1651, published *A Description of New Philosophical Furnaces, or A New Art of*

Distilling, which was a translation from the *Furni Novi Philosophici* and *De Aura Tinctura*, German works with Latin titles by the German chemist Johann Rudolph Glauber. The first editions of the original works seem to have been published in the same year, 1651. From the Italian, Richard Waller translated the *Saggi di Naturali Esperienze* of the Accademia del Cimento of Florence, published in 1667, under the title *Essayes of Natural Experiments*, published in 1684. The work of this short-lived society (1657–67) was very important, laying the foundation of modern physics. Robert Boyle and other British virtuosi repeated the experiments described in the *Saggi* and carried them further. In 1686 William Harris translated the fifth edition of the *Cours de Chymie* of the French savant Nicholas Lemery, published also in 1686, as *A Course of Chymistry*. The treatise *De Arte Vitraria*, by the abbot Antonio Neri, was published as *The Art of Glass*, in 1662, by Christopher Merrett, who also edited a Latin edition published in Amsterdam in 1668.

The increase in the number of translations of works on science after the Civil War, and especially after the Restoration, illustrates the growth of public interest in this branch of learning. The English translations of the Latin works of British authors written in the seventeenth century were usually made after the century had ended, and do not come within our present scope.

GENERAL SCIENCE

Francis Bacon
from *The New Atlantis* (1627)

Description of Salomon's House (Modernized Text)

God bless thee, my son; I will give thee the greatest jewel I have. For I will impart unto thee, for the love of God and men, a relation of the true state of Salomon's House. Son, to make you know the true state of Salomon's House, I will keep this order. First, I will set forth unto you the end of our foundation. Secondly, the preparations and instruments we have for our works. Thirdly, the several employments and functions whereto our fellows are assigned. And fourthly, the ordinances and rites which we observe.

The end of our foundation is the knowledge of causes, and secret motions of things; and the enlarging of the bounds of human empire, to the effecting of all things possible.

The preparations and instruments are these. We have large and deep caves of several depths: the deepest are sunk six hundred fathom; and some of them are digged and made under great hills and mountains: so that if you reckon together the depth of the hill and the depth of the cave, they are (some of them) above three miles deep. For we find that the depth of a hill, and the depth of a cave from the flat, is the same thing; both remote alike from the sun and heaven's beams, and from the open air. These caves we call the Lower Region. And we use them for all coagulations, indurations, refrigerations, and conservations of bodies. We use them likewise for the imitation of natural mines; and the producing also of new artificial metals, by compositions and materials which we use, and lay there for many years. We use them also sometimes (which may seem strange,) for curing of some diseases, and for prolongation of life in some hermits that choose to live there, well accommodated of all things necessary; and indeed live very long; by whom also we learn many things. . . .

We have also perspective-houses, where we make demonstrations of all lights and radiations; and of all colours; and out of things uncoloured and transparent, we can represent unto you all several colours; not in rain-bows, as it is in gems and prisms, but of themselves single. We represent also all multiplications of light, which we carry to great distance, and make so sharp as to discern small points and lines; also all colorations of light: all delusions and deceits of the sight, in figures, magnitudes, motions, colours: all demonstrations of shadows. We find also divers means, yet unknown to you, of producing of light originally from divers bodies. We procure means of seeing objects afar off; as in the heaven and remote places; and represent things near as afar off, and things afar off as near; making feigned distances. We have also helps for the sight, far above spectacles and glasses in use. We have also glasses and means to see small and minute bodies perfectly and distinctly; as the shapes and colours of small flies and worms, grains and flaws in gems, which cannot otherwise be seen; observations in urine and blood, not otherwise to be seen. We make artificial rain-bows, halos, and circles about light. We represent also all manner of reflexions, refractions, and multiplications of visual beams of objects.

We have also precious stones of all kinds, many of them of great beauty, and to you unknown; crystals likewise; and glasses of divers kinds; and amongst them some of metals vitrificated, and other materials besides those of which you make glass. Also a number of fossils, and imperfect minerals, which you have not. Likewise loadstones of prodigious virtue; and other rare stones, both natural and artificial.

We have also sound-houses, where we practise and demonstrate all sounds, and their generation. We have harmonies which you have not, of quarter-sounds, and lesser slides of sounds. Divers instruments of music likewise to you unknown, some sweeter than any you have; together with bells and rings that are dainty and sweet. We represent small sounds as great and deep; likewise great sounds extenuate and sharp; we make divers tremblings and warblings of sounds, which in their original are entire. We represent and imitate all articulate sounds and letters, and the voices and notes of beasts and birds. We have certain helps which set to the ear do further the hearing greatly. We have also divers strange and artificial echos, reflecting the voice many times, and as it were tossing it: and some that give back the voice louder than it came; some shriller, and some deeper; yea, some rendering the voice differing in the letters or articulate sound from that they receive. We have also means to convey sounds in trunks and pipes, in strange lines and distances.

We have also perfume-houses; wherewith we join also practices of taste. We multiply smells, which may seem strange. We imitate smells, making all smells to breathe out of other mixtures than those that give them. We make divers imitations of taste likewise, so that they will deceive any man's taste. And in this house we contain also a confiture-house; where we make all sweet-meats, dry and moist, and divers pleasant wines, milks, broths, and sallets, far in greater varietv than you have.

We have also engine-houses, where are prepared engines and instruments for all sorts of motions. There we imitate and practise to make swifter motions than any you have, either out of your muskets or any engine that you have, and to make them and multiply them more easily, and with small force, by wheels and other means: and to make them stronger, and more violent than yours are; exceeding your greatest cannons and basilisks. We represent also ordnance and instruments of war, and engines of all kinds: and likewise new mixtures and compositions of gun-powder, wildfires burning in water, and unquenchable. Also fire-works of all variety both for pleasure and use. We imitate also flights of birds; we have some degrees of flying in the air; we have ships and boats for going under water, and brooking of seas; also swimming-girdles and supporters. We have divers curious clocks, and other like motions of return, and some perpetual motions. We imitate also motions of living creatures, by images of men, beasts, birds, fishes, and serpents. We have also a great number of other various motions, strange for equality, fineness, and subtilty.

We have also a mathematical house, where are represented all instruments, as well of geometry as astronomy, exquisitely made.

We have also houses of deceits of the senses; where we represent all manner of feats of juggling, false apparitions, impostures, and illusions; and their fallacies. And surely you will easily believe that we that have so many things truly natural which induce admiration, could in a world of particulars deceive the senses, if we would disguise those things and labour to make them seem more miraculous. But we do hate all impostures and lies: insomuch as we have severely forbidden it to all our fellows, under pain of ignominy and fines, that they do not shew any natural work or thing, adorned or swelling; but only pure as it is, and without all affectation of strangeness.

These are, my son, the riches of Salomon's House.

For the several employments and offices of our fellows; we have twelve that sail into foreign countries, under the names of other nations, for our own we conceal, who bring us the books, and abstracts, and patterns of experiments of all other parts. These we call Merchants of Light.

We have three that collect the experiments which are in all books. These we call Depredators.

We have three that collect the experiments of all mechanical arts; and also of liberal sciences; and also of practices which are not brought into arts. These we call Mystery-men.

We have three that try new experiments, such as themselves think good. These we call Pioneers or Miners.

We have three that draw the experiments of the former four into titles and tables, to give the better light for the drawing of observations and axioms out of them. These we call Compilers.

We have three that bend themselves, looking into the experiments of their fellows, and cast about how to draw out of them things of use and practice for man's life, and knowledge as well for works as for plain demonstration of causes, means of natural divinations, and the easy and clear discovery of the virtues and parts of bodies. These we call Dowry-men or Benefactors.

Then after divers meetings and consults of our whole number, to consider of the former labours and collections, we have three that take care, out of them, to direct new experiments, of a higher light, more penetrating into nature than the former. These we call Lamps.

We have three others that do execute the experiments so directed, and report them. These we call Inoculators.

Lastly, we have three that raise the former discoveries by experiments into greater observations, axioms, and aphorisms. These we call Interpreters of Nature.

We have also, as you must think, novices and apprentices, that the succession of the former employed men do not fail; besides a great number of servants and attendants, men and women. And this we do also; we have consultations, which of the inventions and experiences which we have discovered shall be published, and which not: and take all an oath of secrecy, for the concealing of those which we think fit to keep secret: though some of those we do reveal sometimes to the state, and some not.

Thomas Sprat:
from *The History of the Royal Society* (1667)

The Work of Christopher Wren

In the whole progress of this Narration, I have been cautious to forbear commending the labours of any Private Fellows of the Society. For this, I need not make any Apology to them; seeing it would have been an inconsiderable Honour, to be prais'd by so mean a Writer: But now I must break this Law, in the particular case of Dr Christopher Wren: For doing so, I will not alledge the excuse of my Friendship to him: though that perhaps were sufficient; and it might well be allow'd me to take this occasion of Publishing it: But I only do it on the meer consideration of Justice: For in turning over the Registers of the Society, I perceived that many excellent things, whose first invention ought to be ascrib'd to him, were casually omitted: This moves me to do him right by himself, and to give this separate Account of his indeavours, in promoting the Design of the Royal Society, in the small time wherein he has had the opportunity of attending it.

The first instance I shall mention, to which he may lay peculiar claim, is the Doctrine of Motion, which is the most considerable of all others, for establishing the first Principles of Philosophy, by Geometrical Demonstrations. This Des Cartes had before begun, having taken up some Experiments of this kind upon Conjecture, and made them the first Foundation of his whole System of Nature: But some of his Conclusions seeming very questionable, because they were only deriv'd from the gross Trials of Balls meeting one another at Tennis, and Billiards: Dr Wren produc'd before the Society, an Instrument to represent the effects of all sorts of Impulses, made between two hard globous Bodies, either of equal, or of different bigness, and swiftness, following or meeting each other, or the one moving, the other at rest. From these varieties arose many unexpected effects; of all which he demonstrated the true Theories, after they had been confirm'd by many hundreds of Experiments in that Instrument. These he propos'd as the Principles of all Demonstrations in Natural Philosophy: Nor can it seem strange, that these Elements should be of such Universal use; if we consider that Generation, Corruption, Alteration, and all the Vicissitudes of Nature, are nothing else but the effects arising from the meeting of little Bodies, of differing Figures, Magnitudes, and Velocities.

The Second Work which he has advanc'd, is the History of Seasons: which will be of admirable benefit to Mankind, if it shall be constantly pursued, and deriv'd down to Posterity. His proposal therefore was, to comprehend a Diary of Wind, Weather, and other conditions of the Air, as to Heat, Cold, and Weight; and also a General Description of the Year, whether contagious or healthful to Men or Beasts; with an Account of Epidemical Diseases, of Blasts, Mill-dews, and other accidents, belonging to Grain, Cattle, Fish, Fowl, and Insects. And because the difficulty of a constant Observation of the Air, by Night and Day, seem'd invincible, he therefore devis'd a Clock to be annex'd to a Weather-Cock, which mov'd a Rundle cover'd with Paper, upon which the Clock mov'd a Black-lead-Pencil; so that the Observer by the Traces of the Pencil on the Paper, might certainly conclude, what Winds had blown in his absence, for twelve hours space: After a like manner he contriv'd a Thermometer to be its own Register; And because the usual Thermometers were not found to give a true measure of the extension of the Air, by reason that the accidental gravity of the liquor, as it lay higher or lower in the Glass, weigh unequally on the Air, and gave it a farther contraction or extension, over and above that which was produc'd by heat and cold; therefore he invented a Circular Thermometer, in which the liquor occasions no fallacy, but remains always in one height moving the whole Instrument, like a Wheel on its Axis.

He has contriv'd an Instrument to measure the quantities of Rain that falls: This as soon as it is full, will pour out itself, and at the year's end discover how much Rain has fallen on such a space of Land, or other hard superficies, in order to the Theory of Vapours, Rivers, Seas, &c.

He has devis'd many subtil ways for the easier finding the gravity of the Atmosphere, the degrees of drought and moisture, and many of its other accidents. Amongst these Instruments there are Balances which are useful to other purposes, that shew the weight of the Air by their spontaneous inclination.

Amongst the new Discoveris of the Pendulum, these are to be attributed to him, that the Pendulum in its motion from rest to rest; that is, in one descent and ascent, moves unequally in equal times, according to a line of sines: That it would continue to move either in Circular, or Eliptical Motions; and such Vibrations would have the same Periods with those that are reciprocal; and that by a complication of several Pendulums depending one upon another, there might be represented motions like the planetary Helical Motions, or more intricate: And yet that these Pendulums would discover without confusion (as the Planets do) three or four several Motions, acting upon one body with differing Periods; and that there may be produc'd a Natural standard for Measure from the Pendulum for vulgar use,

He has invented many ways to make Astronomical Observations more accurate and easy: He has fitted and hung Quadrants, Sextants, and Radii, more commodiously than formerly: He has made two Telescopes, to open with a joynt like a Sector, by which Observers may infallibly take a distance to half minutes, and find no difference in the same Observation reiterated several times; nor can any warping or luxation of the Instrument hinder the truth of it.

He has added many sorts of Retes, Screws, and other devises to Telescopes, for taking small distances and apparent Diameters to Seconds. He has made apertures to take in more or less light, as the Observer pleases, by opening and shutting like the Pupil of the Eye, the better to fit Glasses to Crepusculine Observations: He has added much to the Theory of Dioptrics; much to the Manufacture it self of grinding good Glasses. He has attempted, and not without success, the making of Glasses of other forms than Spherical: He has exactly measured and delineated the Spheres of the Humours in the Eye, whose proportions one to another were only guess'd at before. This accurate discussion produc'd the Reason, why we see things erected, and that Reflection conduces as much to Vision as Refraction.

He discours'd to them a natural and easy Theory of Refraction, which exactly answer'd every Experiment. He fully demonstrated all Dioptrics in a few Propositions, shewing not only (as in Kepler's Dioptrics) the common Properties of Glasses, but the Proportions by which the individual Rays cut the Axis, and each other; upon which the Charges (as they are usually called) of Telescopes, or the Proportion of the Eye-glasses and Apertures are demonstrably discover'd.

He has made constant Observations on Saturn; and a Theory of that Planet, truly answering all Observations, before the printed Discourse of Hugonius on that Subject appear'd.

He has essay'd to make a true Selenography by measure; the World having nothing yet but Pictures, rather than Surveys and Maps of the Moon. He has stated the Theory of the Moon's Libration, as far as his Observations could carry him. He has compos'd a Lunar Globe, representing not only the Spots, and various degrees of whiteness upon the Surface, but the Hills, Eminencies, and Cavities moulded in solid Work. The Globe thus fashioned into a true Model of the Moon, as you turn it to the Light, represents all the Menstrual phases, with the variety of Appearances that happen from the Shadows of the Mountains and Valleys. He has made Maps of the Pleiades, and other Telescopical Stars; and propos'd Methods to determine the great doubt of the Earth's motion or rest, by the small Stars about the Pole to be seen in large Telescopes.

In order to Navigation he has carefully pursu'd many Magnetical Experiments; of which this is one of the noblest and most fruitful Speculation: A large Terella is plac'd in the midst of a Plane Board, with a hole into which the Terella is half immers'd, till it be like a Globe, with the Poles in the Horizon. Then is the Plane dusted over with steel-filings equally from a Sieve: The Dust by the Magnetical virtue is immediately figur'd into Furrows that bend like a sort of Helix, proceeding as it were out of one Pole, and returning into the other: And the whole Plane is thus figur'd like the Circles of a Planisphere.

Prospects of Scientific Discovery in America and Elsewhere

This is the most natural Method of the Foundation and Progress of Manual Arts. And they may still be advanc'd to a higher Perfection, than they have yet obtain'd, either by the Discovery of new Matter, to imploy Mens Hands, or by a new Transplantation of the same Matter, or by handling the old Subjects of Manufactures after a new way, in the same Places.

And first, we have reason to expect, that there may still arise new Matter to be manag'd by Human Art and Diligence; and that from the parts of the Earth that are yet unknown, or from the new discover'd America, or from our own Seas and Land, that have been long search'd into, and inhabited.

If ever any more Countries, which are now hidden from us, shall be reveal'd, it is not to be question'd, but there will be also opened to our Observation, very many kinds of living Creatures, of Minerals, of Plants, nay, of Handicrafts, with which we have been hitherto unacquainted. This may well be expected, if we remember, that there was never yet any Land discover'd, which has not given us divers new sorts of Animals, and Fruits of different Features and Shapes, and Virtues from our own, or has not supplied us with some new artificial Engine, and Contrivance.

And that our Discoveries may still be inlarg'd to farther Countries, it is a good Proof, that so many spacious Shores and Mountains, and Promontories, appear to our Southern and Northern Sailors; of which we have yet no Account, but only such as could be taken by a remote Prospect at Sea. From whence, and from the Figure of the Earth, it may be concluded, that almost as much space of Ground remains still in the Dark, as was fully known in the times of the Assyrian or Persian Monarchy. So that without assuming the vain prophetick Spirit, which I lately condemn'd, we may foretel, that the Discovery of another new World is still behind.

To accomplish this, there is only wanting the Invention of Longitude, which cannot now be far off, seeing it is generally allow'd to be feasible, seeing so many Rewards are ready to be heap'd on the Inventors; and (I will also add) seeing the Royal Society has taken it into its peculiar care. This, if it shall be once accomplish'd, will make well-nigh as much alteration in the World, as the Invention of the Needle did before: And then our Posterity may outgo us, as much as we can travel farther than the Antients; whose Demy Gods and Heroes did esteem it one of their chief Exploits, to make a Journey as far as the Pillars of Hercules. Whoever shall think this to be a desperate Business, they can only use the same Arguments, wherewith Columbus was at first made ridiculous, if he had been discourag'd by the Raillery of his Adversaries, by the Judgment of most Astronomers of his time, and even by the Intreaties of his own companions; but three Days before he had a sight of Land, we had lost the Knowledge of half the World at once.

And as for the new discover'd America, 'tis true, that has not been altogether useless to the Mechanic Arts: But still we may guess, that much more of its Bounty is to come, if we consider, that it has not yet been shewn above two hundred Years; which is scarce enough time to travel it over, describe, and measure it, much less to pierce into all its Secrets. Besides this, a good part of this Space was spent in the Conquest and settling the

Spanish Government, which is a Season improper for Philosophical Discoveries. To this may be added, that the chief Design of the Spaniards thither, has been the Transportation of Bullion; which being so profitable, they may well be thought to have overseen many other of its Native Riches. But above all, let us reflect on the Temper of the Spaniards themselves: They suffer no Strangers to arrive there: they permit not the Natives to know more than becomes their Slaves. And how unfit the Spanish humour is to improve Manufactures, in a Country so distant as the West-Indies, we may learn by their Practice in Spain itself, where they commonly disdain to exercise any Manual Crafts, and permit the Profit of them to be carried away by Strangers.

From all this we may make this Conclusion, That if ever that vast Tract of Ground shall come to be more familiar to Europe, either by a free Trade, or by Conquest, or by any other Revolution in its Civil Affairs, America will appear quite a new Thing to us; and may furnish us with an abundance of Rarities, both Natural and Artificial; of which we have been almost as much depriv'd by its present Masters, as if it had still remain'd a part of the unknown World.

But lastly, to come nearer home, we have no ground to despair, but very much more Matter, which has been yet unhandled, may still be brought to light, even in the most civil and most peopled Countries; whose Lands have been thoroughly measur'd by the Hands of the most exact Surveyors; whose under-ground Riches have been accurately pry'd into; whose Cities, Islands, Rivers, and Provinces, have been describ'd by the Labours of Geographers. It is not to be doubted, but still there may be an infinite number of Creatures over our Heads, round about us, and under our Feet, in the large Space of the Air, in the Caverns of the Earth, in the Bowels of Mountains, in the Bottom of Seas, and in the Shades of Forests, which have hitherto escap'd all mortal Senses. In this the Microscope alone is enough to silence all Opposers. Before that was invented, the chief help that was given to the Eyes by Glasses, was only to strengthen the dim Sight of old Age; but now by the means of that excellent Instrument, we have a far greater Number of different kinds of Things reveal'd to us, than were contain'd in the visible Universe before; and even this is not yet brought to Perfection: The chief Labours that are publish'd in this way, have been the Observations of some Fellows of the Royal Society, nor have they as yet apply'd it to all Subjects, nor tried it in all Materials and Figures of Glass.

To the Eyes therefore there may still be given a vast addition of Objects: And proportionably to all the other Senses. This Mr Hook has undertaken to make out, that Tasting, Touching, Smelling and Hearing, are as improveable as the Sight; and from his excellent Performances in the one, we may well rely on his Promise in all the rest.

Mechanical Inventions

For it may be observ'd, that the greatest part of all our New Inventions have not been raised from Subjects before untouch'd (though they also have given us very many) but from the most studied and most familiar Things, that have been always in Mens Hands and Eyes. For this I shall only instance in Printing, in the Circulation of the Blood, in Mr Boyle's Engine for the sucking out of Air, in the making of Guns, in the Microscopical Glasses, and in the Pendulum Clocks of Hugenius. . . .

And yet in all these the most obvious Things, the greatest Changes have been made by late Discoveries; which cannot but convince us, that many more are still to come from Things that are as common, if we shall not be wanting to ourselves.

And this we have good reason to trust will be effected, if this Mechanic Genius, which now prevails in these Parts of Christendom, shall happen to spread wider amongst ourselves, and other Civil Nations; or if by some good Fate it shall pass farther on to other Countries that were yet never fully civiliz'd. We now behold much of the Northern Coasts of Europe and Asia, and almost all Afric, to continue in the rude State of Nature: I wish I had not an Instance nearer Home, and that I did not find some Parts of our own Monarchy in as bad a Condition. But why may we not suppose, that all these may in course of Time be brought to lay aside the untam'd Wildness of their present Manners? Why should we use them so cruelly as to believe, that the Goodness of their Creator has not also appointed them their Season of polite and happy Life, as well as us? Is this more unlikely to happen, than the Change that has been made in the World these last seventeen hundred Years? This has been so remarkable, that if Aristotle, and Plato, and Demosthenes, should now arise in Greece again, they would stand amaz'd at the horrible Devastation of that which was the Mother of Arts. And if Cæsar and Tacitus should return to Life, they would scarce believe this Britain, and Gaul, and Germany, to be the same which they describ'd: They would now behold them cover'd over with Cities and Palaces, which were then over-run with Forests and Thickets: They would see all manner of Arts flourishing in these Countries, where the chief Art that was practis'd in their Time, was that barbarous one of painting their Bodies, to make them look more terrible in Battle.

This then being imagin'd, that there may some lucky Tide of Civility flow into those Lands, which are yet savage, there will a double Improvement thence arise, both in respect of ourselves and them: For even the present skilful Parts of Mankind, will be thereby made more skilful; and the other will not only increase those Arts which we shall bestow upon them, but will also venture on new Searches themselves.

If any shall doubt of the first of these Advantages, let them consider that the spreading of Knowledge wider, does beget a higher and a clearer Genius in those that enjoy'd it before.

But the chief Benefit will arise from the New Converts;

for they will not only receive from us our Old Arts, but in their first Vigour will proceed to new ones that were not thought of before. This is reasonable enough to be granted: For seeing they come fresh and unwearied, and the Thoughts of Men being most violent in the first opening of their Fancies; it is probable they will soon pass over those Difficulties about which these People, that have been long Civil, are already tir'd. To this purpose I might give as many Examples as there have been different Periods of civilizing; that those Nations which have been taught, have prov'd wiser and more dextrous than their Teachers. The Greeks took their first Hints from the East; but out-did them in Music, in Statuary, in Graving, in Limning, in Navigation, in Horsemanship, in Husbandry, as much as the Ægyptians or Assyrians exceed their unskilful Ancestors in Architecture, Astronomy, or Geometry. The Germans, the French, the Britains, the Spaniards, the modern Italians, had their Light from the Romans; but surpass'd them in most of their own Arts, and well nigh doubled the ancient Stock of Trades deliver'd to their keeping.

So then, the whole Prize is not yet taken out of our Hands: The Mechanic Invention is not quite worn away, nor will be, as long as new Subjects may be discovered, as long as our old Materials may be alter'd or improv'd, and as long as there remains any Corner of the World without Civility. Let us next observe, whether Men of different ways of Life are capable of performing any Thing towards it, besides the Artificers themselves. This will quickly appear undeniable, if we will be convinc'd by Instances; for it is evident, that diverse sorts of Manufactures have been given us by Men who were not bred up in Trades that resembled those which they discovered. I shall mention Three; that of Printing, Powder, and the Bow-Dye. The admirable Art of Composing Letters, was so far from being started by a Man of Learning, that it was the Device of a Soldier: And Powder (to make Recompence) was invented by a Monk, whose course of Life was most averse from handling the Materials of War. The ancient Tyrian Purple was brought to light by a Fisher; and if ever it shall be recover'd, it is likely to be done by some such Accident. The Scarlet of the Moderns is a very beautiful Colour; and it was the Production of a Chymist, and not of a Dyer.

And indeed the Instances of this kind are so numerous, that I dare in general affirm, That those Men who are not peculiarly conversant about any one sort of Arts, may often find out their Rarities and Curiosities sooner, than those who have their Minds confin'd wholly to them. If we weigh the Reasons why this is probable, it will not be found so much a Paradox, as perhaps it seems at the first Reading. The Tradesmen themselves, having had their Hands directed from their Youth in the same Methods of Working, cannot when they please so easily alter their Custom, and turn themselves into new Roads of Practice. Besides this, they chiefly labour for present Livelihood, and therefore cannot defer their Expectations so long, as is commonly requisite for the ripening of any new Contrivance. But especially having long handled their Instruments in the same Fashion, and regarded their Materials with the same Thoughts, they are not apt to be surpriz'd much with them, nor to have any extraordinary Fancies, or Raptures about them.

These are the usual Defects of the Artificers themselves: Whereas the Men of freer Lives, have all the contrary Advantages: They do not approach those Trades, as their dull and unavoidable, and perpetual Employments, but as their Diversions: They come to try those Operations, in which they are not very exact, and so will be more frequently subject to commit Errors in their Proceeding: Which very Faults and Wandrings, will often guide them into new Light, and new Conceptions: And lastly, there is also some Privilege to be allow'd to the Generosity of their Spirits, which have not been subdu'd, and clogg'd by any constant Toil, as the others. Invention is an Heroic Thing, and plac'd above the reach of a low and vulgar Genius: It requires an active, a bold, a nimble, a restless Mind:

A thousand Difficulties must be contemn'd, with which a mean Heart would be broken; many Attempts must be made to no Purpose; much Treasure must sometimes be scatter'd without any Return; much Violence and Vigour of Thoughts must attend it; some Irregularities and Excesses must be granted it, that would hardly be pardon'd by the severe Rules of Prudence. All which may persuade us, that a large and an unbounded Mind is likely to be the Author of greater Productions, than the calm, obscure, and fetter'd Endeavours of the Mechanics themselves: And that as in the Generation of Children, those are usually observ'd to be most sprightly, that are the stolen Fruits of an unlawful Bed: so in the Generations of the Brains, those are often the most vigorous and witty, which Men beget on other Arts, and not on their own.

ISAAC NEWTON:

Letter to Henry Oldenburg (January 25, 1675/6)

SIR,

I received both yours, and thank you for your care in disposing those things between me and Mr Linus. I suppose his friends cannot blame you at all for printing his first letter, it being written, I believe, for that end, and they never complaining of the printing of that, but of the not printing that, which followed, which I take myself to have been *per accidens* the occasion of, by refusing to answer him. And though I think I may truly say, I was very little concerned about it, yet I must look upon it as the result of your kindness to me, that you was unwilling to print it without an answer.

As to the paper of Observations, which you move in the name of the Society to have printed, I cannot but return them my hearty thanks for the kind acceptance they meet with there, and know not how to deny any thing, which they desire should be done. Only I think it will be best to suspend the printing of them for a while, because I have some thoughts of writing such another set of Observations for determining the manner of the productions of colours by the prism, which, if done at all, ought to precede that now in your hands, and will do best to be joined with it. But this I cannot do presently, by reason of some incumbrances lately put upon me by some friends, and some other business of my own, which at present almost take up my time and thoughts.

The additions, that I intended, I think I must, after putting you to so long expectations, disappoint you in; for it puzzles me how to connect them with what I sent you; and if I had those papers, yet I doubt the things I intended will not come in so freely as I thought they might have done. I could send them described without dependance on those papers; but I fear I have already troubled your Society and yourself too much with my scribbling, and so suppose it may do better to defer them till another season. I have therefore at present only sent you two or three alterations, though not of so great moment, that I need have staid you for them; and they are these:

Where I say, that the frame of nature may be nothing but æther condensed by a fermental principle, instead of these words write, that it may be nothing but various contextures of some certain ætherial spirits or vapours condensed as it were, by precipitation, much after the manner, that vapours are condensed into water, or exhalations into grosser substances, though not so easily condensable; and after condensation wrought into various forms, at first by the immediate hand of the Creator, and ever since by the power of nature, who by virtue of the command, Increase and multiply, became a complete imitator of the copies set her by the Protoplast. Thus perhaps may all things be originated from æther, &c.

A little after, when I say, the ætherial spirit may be condensed in fermenting or burning bodies, or otherwise inspissated in the pores of the earth to a tender matter, which may be, as it were, the *succus nutritius* of the earth, or primary substance, out of which things generable grow: instead of this you may write, that that spirit may be condensed in fermenting or burning bodies, or otherwise coagulated in the pores of the earth and water into some kind of humid active matter, for the continual uses of nature, adhering to the sides of those pores after the manner that vapours condense on the sides of a vessel.

In the same paragraph there is, I think, a parenthesis, in which I mention volatile salt-petre. Pray strike out that parenthesis, lest it should give offence to somebody.

Also where I relate the experiment of little papers made to move variously with a glass rubbed, I would have all that struck out, which follows about trying the experiment with leaf-gold.

Sir, I am interrupted by a visit, and so must in haste break off.

Yours,

Jan. 25, 1675/6 ISAAC NEWTON

BOTANY

John Ray:

from *Philosophical Letters between the late learned Mr Ray and Several of his Ingenious Correspondents, Natives and Foreigners* (1718)

Of the Number of Plants

To determine precisely what Number of Plants there are in the World, is a thing, if not absolutely, at least morally impossible, as we shall prove anon. But before we make any conjecture about their Number, it will be requisite to debate these two Questions. 1. Whether there have been or are yearly any new Species produced besides what were at first created? 2. Whether there have been, or may be any Species lost or destroyed? For if either of these be affirmed, in vain would it be to enquire the number of the Plants; it being uncertain, and variable every Year, and that possibly to a very great excess, or defect. For the causes of these Destructions and Productions being accidental, there is no reason why one should exactly, or in any near proportion, balance and compensate the other.

Of the first Question, those that hold the affirmative for proof of their Opinion, alledge common Experience: For doth not every new Year afford us new sorts of Flowers and Fruits? And consequently new sorts of Plants? Are not our Gardens and Orchards yearly enriched with new sorts, for Example, of July Flowers, Tulips, and Anemonies, of Apples and Pears? Do not our Gardeners sell us these for distinct Species? And do not Herbarists generally enumerate and describe them for such? What Herbal doth not make, for instance, *Caryophyllus*, or *Viola*, *Paralysis*, with a double Flower, different kinds from those with a single? I Answer, It is true, they do so; but if we examine and consider wherein their Differences consist, we shall find reason to doubt whether they be specifically distinct or no; nay rather to conclude, they are not. First, as for Flowers. The main, if not only difference between these pretended new Species and the old, we shall find to consist either in the colour of the Flower, or the multiplicity of its Leaves. Now that neither of these is sufficient to infer a specifical Difference, is, I think, evident, unless we will admit that an European, and an Ethiopian, are two Species of Men, because one is black and the other white; or an European and an Indian, because the one hath a thick Beard, and the other none at all, or but a few straggling Hairs instead of it: The whole diversity being induced by the Climate, or Soil, or Nourishment, as in other Animals, is manifest. First, what effect the plenty and diversity of Food, and different manner of Living hath, appears in domestick Animals, *ex. gr.* Swine, Ducks, Geese, &c. which do frequently vary their Colours; whereas the wild of those kinds retain constantly the same; and not their Colours only, but the tastes of their Flesh, it requiring no very critical Palate to distinguish between the Flesh of tame and wild Beasts, or Fowl. As for the Colour, though wild Animals taken and brought up tame, do not usually themselves *in individuo* change their Colours, but after two or three Generations their Breed; yet sometimes they do, as I myself have seen a Bull-finch, which kept in a Cage, after some Years, from the usual Colour of that Bird, turn'd to be Cole black. 2. What influence the diversity of Soil and Climate hath upon divers Animals, as to the altering their Colour, and other Accidents, appears in divers Instances. From the difference of Climate, or constant inspection of Snow, it proceeds that in the Alps and other high Mountains, and also in those cold Northern Countries where the Earth, for more than half the Year, is continually cover'd with Snow, there are found many Animals white of those sorts, which are usually and naturally of another Colour; as for Example, white

Bears, white Foxes, white Hares, white Ravens, white Blackbirds; and many others, as I myself have seen in Italy. That Hares upon the Alps; and in the cold Northern Regions, do in the Winter time, change their Colour to White, and in the Summer again return to their usual and natural Colour; though I find it deliver'd by good Authors, and attested by credible Persons, I dare not peremptorily assert: But that the influence of the Soil and Climate is great, appears farther in our Lancashire and Sussex Beasts; of which the former have fair, large, and well-spread Horns, the latter small and crooked; and if into Sussex you translate these Cattel out of Lancashire, their Race by degrees will degenerate, and come to the Shape of the Natives. So we see the Horses in Flanders have large and hairy Pasterns, which the English Breed have not: And it is reported for a Truth, that there is a Pasture upon a Hill call'd Haselbedge, in the Peak of Derbyshire, near little Hucklow, which will turn the Hair of Kine that feed thereupon, to a grey Colour in three Years space. Now if diversity of Soil, Food, Climate, or other external Circumstances, breed such variety and difference among Animals of the same Species, much more then may it among Plants, which are less free in the choice of their Nourishment, and constantly affix'd to the Place where they chance to spring up.

NEHEMIAH GREW:
from *The Anatomy of Plants* (1682)

An Idea of a Philosophical History of Plants

IF we take an account of the Degrees whereunto the Knowledge of Vegetables is Advanced, it appeareth, That besides the great Varieties, which the Successful Arts of Florists, or Transplantations from one Climate to another, have produced; we have very many Species brought to light, especially Natives of the Indies, which the Ancients, for any thing that appears in their Writings now extant, were ignorant of. In which particular Clusius, Columna, Bauhinus, Boccone, and others, have performed much. Withall, That their Descriptions (of all Parts above ground) their Places and Seasons, are with good diligence and preciseness set before us. Likewise their Order and Kindred: for the adjusting whereof our Learned Countryman Mr Ray, and Dr Morrison, have both taken very laudable pains. As also the ordering of them with respect to their Alimental and Mechanick Uses; for which, amongst others, Mr Evelyn and Dr Beal have deserved many thanks, and great praise. We are also informed, of the Natures and infallible Faculties of many of them. Whereunto so many as have assisted, have much obliged their Posterity.

2. By due Reflection upon what hath been Performed; it also appears, what is left Imperfect and what Undone. For the Virtues of most Plants, are with much uncertainty, and too promiscously ascribed to them. So that if you turn over an Herbal, you shall find almost every Herb, to be good for every Disease. And of the Virtues of many, they are altogether silent. And although, for the finding out, and just appropriation of them, they have left us some Rules, yet not all. The Descriptions likewise of many, are yet to be perfected; especially as to their Roots. Those who are very curious about the other Parts, being yet here too remiss. And as for their Figures, it were much to be wished, That they were all drawn by one Scale; or, at most, by Two; one, for Trees and Shrubs; and another for Herbs. Many likewise of their Ranks and Affinities, are yet undetermined. And a great number of Names, both English and Latine, not well given. So what we call Goat's Rue, is not at all of kin to that Plant, whose Generical Name it bears. The like may be said of Wild-Tansy, Stock-July-Flowers, Horse-Radish, and many more. So also when we say Bellis Major, and Minor, as we commonly do, these Names would intimate, That the Plants to which they are given, differ (as the great double Marigold, doth from the less) only in Bulk: whereas, in truth, they

are two Species of Plants. So we commonly say, Centaurium Majus and Minus, Chelidonium Majus and Minus, and of others in like manner, which yet are distinct Species, and of very different tribes. But for the Reason of Vegetation, and the Causes of all those infinite Varieties therein observable (I mean so far as Matter, and the various Affections hereof, are instrumental thereunto) almost all Men have seemed to be unconcerned.

3. That Nothing hereof remaineth further to be known, is a Thought not well Calculated. For if we consider how long and gradual a Journey the Knowledge of Nature is; and how short a Time we have to proceed therein; as on the one hand, we shall conclude it our ease and profit, To see how far Others have gone before us: so shall we beware on the other, That we conceive not unduly of Nature, whilst we have a just value for Those, who were but her Disciples, and instructed by Her. Their Time and Abilities both, being short to her; which, as She was first Designed by Divine Wisdom; so may Her vast Dimensions best be adjudged of, in being compared therewith. It will therefore be our Prudence, not to insist upon the Invidious Question, Which of Her Scholars have taken the fairest measure of Her; but to be well satisfied, that as yet She hath not been Circumscribed by Any.

4. Nor doth it more behove us to consider, how much of the Nature of Vegetation may lie before us yet unknown; Than, to believe, a great part thereof to be knowable. Not concluding from the acknowledged, much less supposed Insuccessfulnes, of any Mens Undertakings: but from what may be accounted Possible, as to the Nature of things themselves; and from Divine Providence, by Infinite Ways conducting to the Knowledge of them. Neither can we determine how great a part This may be: Because, It is impossible to Measure, what we See not. And since we are most likely to under-measure, we shall hereby but intrench our Endeavours, which we are not wont to carry beyond the Idea, which we have of our Work.

5. And how far soever this kind of Knowledge may be attainable its being so far also worthy our attainment will be granted. For beholding the Many and Elegant Varieties, wherewith a Field or Garden is adorned; Who would not say, That it were exceeding pleasant to know what we See: and not more delightful, to one who has Eyes, to discern that all is very fine; than to another who hath Reason, to understand how. This surely were for a Man to take a True Inventory of his Goods, and his best way to put a price upon them. Yea it seems, that this were not only to be Partaker of Divine Bounty; but also, in some degree, To be Copartner in the Secrets of Divine Art. That which were very desireable, unless we should think it impertinent for us to design the Knowing of That, which God hath once thought fit to Do.

6. If for these, and other Reasons, an inquiry into the Nature of Vegetation may be of good Import; It will be requisite, to see, first of all, What may offer itself to be enquired of; or to understand, what our Scope is: That so doing, we may take our aim the better in making, and having made, in applying our Observations thereunto. Amongst other Inquiries therefore, such as these deserve to be proposed. First, by what means it is that a Plant, or any Part of it, comes to Grow, a Seed to put forth a Root and Trunk; and this, all the other Parts, to the Seed again; and all these being formed, by continual Nutrition still to be increased. How the Aliment by which a Plant is fed, is duly prepared in its several Parts; which way it is conveyed unto them; and in what manner it is assimilated to their respective Natures in them all. Whence this Growth and Augmentation is not made of one, but many differing Degrees, unto both extremes of small and great; whether the comparison be made betwixt several Plants, or the several Parts of one. How not only their Sizes, but also their Shapes are so exceeding various; as of Roots, in being Thick or Slender, Short or Long, Entire or Parted, Stringed or Ramified, and the like: of Trunks, some being more Entire,

others Branched, others Shrub'd: of Leaves which are Long or Round, Even-edg'd or Escallop'd, and many other ways different, yet always Flat: and so for the other Parts. Then to inquire, What should be the reason of their various Motions; that the Root should descend; that its descent should sometimes be perpendicular, sometimes more level: That the Trunk doth ascend; and that the ascent thereof, as to the space of Time wherein it is made, is of different measures: and of divers other Motions, as they are observable in the Roots, Trunks, and other Parts of Plants. Whence again, these Motions have their Different and Stated Terms; that Plants have their set and peculiar Seasons for their Spring or Birth, for their Full Growth, and for their Teeming; and the like. Further, what may be the Causes as of the Seasons of their Growth; so of the Periods of their Lives; some being Annual, others Biennial, others Perennial; some Perennial both as to their Roots and Trunks; and some as to their Roots only. Then, as they pass through these several Seasons of their Lives, in what manner their convenient feeding, housing, cloathing or protection otherwise, is contrived; wherein, in this kind and harmonious Œconomy, one Part, may be officious to another, for the preservation of the health and life of the whole. And lastly, what care is taken, not only for themselves, but for their Posterity; in what manner the Seed is prepared, formed and fitted for Propagation: and this being of so great concernment, how sometimes the other Parts also, as Roots, in putting forth Trunks; Trunks in putting forth Roots; yea in turning oftentimes into Roots themselves; whereof, in the Second Book of the Anatomy of Plants, I shall give some instances. With other Heads of Inquiry of this kind.

Martin Lister:
from the *Philosophical Transactions of the Royal Society* (before 1700)

The Nature and Difference of the Juices of Plants

We observe that mostly Juices of Plants coagulate, whether they be such as are drawn from the Wounds of a Plant, or such as do spontaneously exsudate: And yet that Exsudation seems to be often accidental too, that is, by a Cancer, or some other such like Chance.

And yet I am uncertain what to think of the small purple Blebs and Veins to be observed, more or less on all the *Hypericum* Kind, and on the Threads of the Flower, and the Hairs which cover the Leaves of *Rorella* in like manner. I doubt much, whether this may properly be called an exsudated and coagulated Juice, or no. Our Observations of those of this Tribe, are what follow:

The small green Leaves, next encompassing the yellow Flowers of *Androsæmum Hypericoides Ger.* are set with very small round Blebs, full of a purple Juice; as are likewise, but with two or three only, the very Points or Tops of the yellowish Leaves themselves: Yet the Stalk cut doth not to the Eye discover any such distinct Vessels, carrying that purple Liquor, which makes me suspect it is separated by Coagulation from the rest of the Juice, and reserved in those small Bags.

Hypericum Ger. The purple Juice yielding Blebs, in this Point are upon the Edging, on the outsides of all the Leaves. Also the Stalk, though round, hath a double Edge, on each Side one; and the Blebs or Bags, though but thinly, are yet observable on these very rising Edges too of the Stalks. As for the yellow Flowers themselves the outmost green Leaves, next and immediately encompassing them, have but few purple Stripes, but the yellow Leaves or Flowers, are edged with small purple Bags on the one Side, and stripped with purple Juice yielding Veins on the other. Lastly, On the very Tops of each Thread in the Flower, is one single purple Bag.

Hypericum Ascyron dictum, Caule quadrangulo J.B. In like manner all the Edges on the outsides of all the Leaves, from one End of the Stalk to the other of this Plant, are very thick set with purple Bags. Also in the Flower, all the Threads have one single Bag on the Top; but the Flowers are yellow Leaves, and the green ones encompassing them, have very few purple Spots or Streaks visible.

Hypericum pulchrum Tragi, J.B. Only the yellow Flower-Leaves, and those green ones which next encompasses them, are thick edged with purple Blebs.

Diverse Parts of the same Plant have diverse Faculties, *V.C.P.A.* I add, that diverse Parts of the same Plant, yield from the same Veins different coloured Juices, *v.g.* The Milk in the Root of *Spondylium Ger.* is of a Brimstone Colour; but in the Stalk white.

Amongst those Juices that coagulate and are clammy, some there are which readily break with the Whey.

In the Middle of July I drew and gathered of the Milk of *Lactuca syl. costa spinosa, C.B.* which it freely and plentifully affords. It springs out of the Wound thick as Cream and ropes, and is white; and yet the Milk which came out of the Wounds, made towards the Top of the Plant, was plainly streaked or mixed with a purple Juice, as though one had dashed or sprinkled Cream with a few Drops of Claret. And indeed the Skin of the Plant thereabouts was purplish also, perhaps with Veins. Again in the Shell I drew it; it turned still yellower and thicker, and by and by curdled, that is, the white and thick caseous Part did separate from a thin purple Whey. So the Blood also of Animals, whilst warm, remains liquid and alike: But so soon as cold it cakes, and has a Serum, or Whey, separated from it. Also the caseous part of the Milk of Animals is glutinous and stringy. Further, this Serum came freely from the other, by squeezing betwixt my Fingers; and the Curds I washed in Spring-water, which became immediately like Cags and tougn (draw this Milk immediately, or let it fall off the Plant, into a Shell of fair Water, or other Menstruum; as Vinegar, *S.V.* Spirit of Vitriol, or Sulphur, &c.) and remained still white and dry. As for the purple Whey, after a Day's Insolation, it stifned and became hard, and was easily formed into Cakes, which Cakes were yet very brittle, and would easily crumble into Powder. About December following I broke one of the Cakes made of the caseous Part of the Milk of this Plant; it then proved very brittle and shined, upon breaking, like Rosin: It was then of a dark-brown Colour: Moreover, it burned with a lasting Flame, like Rosin or Wax; and that being melted by Heat, it would draw out into long tough Strings like Birdlime. On the contrary, the purplish Powder, which was the Whey, if put into the Flame of a Candle, would scarce burn with a Flame at all, but soon be turned into a Coal. Lastly, The purple Powder did taste very bitter; whereas the caseous Part was insipid as Wax.

The Milk which the *Trachelium* kind plentifully yields, is very thick, and presently curdles; the serous part, or Whey, being of a brown Colour. These Juices smell sour, something like the Slices of green Apples, which have been long cut.

The thin Milk of *Tithymallus Helioscopius Ger.* springs freely and plentifully; it is very clammy upon the Fingers; it is very white in drawing; it turns upon a Lancet, of a dark bluish; and indeed it is both of the Colour and Consistence of blue skimmed Milk; made up with Wheat Flower into Cakes, it shews itself greasy or oily, and scarce ever dries; it very hardly breaks or coagulates. I kept some of it pure and unmixed, in little Essence-Bottles, stopped lightly with Cork only; in these it broke in process of time, and the Curds were easily to be formed into Cakes; which Cakes burned with a lasting Flame, and being melted drew forth into Strings like Wax; the Whey was clear like fair Water. This broken Milk in all my Bottles was very corrupt and stinking: But the Cakes I made up of this Juice, with Wheat Flower and a little Gum Arabick, dried well, and kept sweet.

Other clammy Juices there are, which do not let go a Whey when they coagulate, but cake altogether.

I made Cakes of the sole or unmixed Juice of *Sonchus lævis, & asper*, without any Addition, and it parted not with any Whey.

Papaver Rheas Ger. bleeds freely a white Juice, and the Heads or Seed-Vessels, when the Flower is gone, do yet bleed. I observed that in gathering it into Shells, it presently turned its white Colour into a yellow one inclining to an Orange. At first springing it roaped, or was but little clammy, and seemed to be very liquid and dilute; yet it did not part with any Whey, but grew stiff, and is very resinous and oily.

Note, The Milks or Juices of Plants seem to be compounded, and mixed of Liquors of different and perhaps contrary Qualities; so that it is probable, if the caseous part shall be Narcotick, for Example, the Whey may not be so or the one may be hurtful, and the other a good and useful Medicament.

Trogopogon flore luteo J.B. yields a Juice which, upon the first springing from the Wound, is white and thick, but immediately turns yellow, and then redder and redder. It is of no unpleasant Taste; it is something glutinous and oily, and parts not with much, if with any Whey, and therefore it is easily formed into Cakes alone.

Convolvulus major J.B. bleeds freely a white Juice as I experienced in the middle of August; not only the Stalk and Leaves, but the white Flowers also in proportion bleed as plentifully as any part else. This Milk is very sharp.

There is also a Juice of a Saffron Colour, which *Chelidonium majus Ger.* wounded freely affords. This Juice breaks not with a Whey, but is easily formed into Cakes, and stiffens in the Sun: It is thick, and of the consistence of Cream, upon the springing out of the Wound.

There is another very clammy Juice, which is of a golden or yellow Colour, upon drawing; and this the Seed-Vessels of *Centaurium luteum perfoliatum C.B.* in July, and after, even where the Seeds therein contained are turned black and ripe, yield plentifully and freely enough. (These Juices, which the Heads, or Seed-vessels of Plants afford, may be thought of the same Nature with those Juices which the Pulp of Fruits affords, the Pulps of Fruits and these exterior Vessels being parts equivalent; (that is, Apples, for example, are nothing else but the Seed-Vessels of their Kernels;) It is liquid upon the first drawing, and after a while it thickens, parting with no Whey; (*N.B.* I call this coagulating too), and this is of the Colour of Amber; it sticks to one's Fingers, and pulls forth into Threads like Bird-lime; it would never become harder than very soft Wax, and that by being dried in the Shade only; for if ever so little be exposed to the Heat of the Sun or Fire, it straightway became exceeding soft. But as for the Cakes I made up of it and Wheat-flower, them I found in my Cabinet in Winter very hard and firm, and the unmixed Cakes still soft. These burn with an unpleasant Smell; they emit a lasting Flame; they still keep their Amber Colour, and draw out into Threads, in burning like wax.

To this we may add the yellow Juice which the Wounds of *Angelica sativa Park*, yield; it will not harden by Insolation, or long keeping (for I have had an Essence-Bottle of it by me these Two Years) yet I perceive it stiffens, and will draw into Threads.

The next sort of coagulate or clammy Juice we have taken notice of, are Gums; and some of them seem long to abide liquid, and perhaps inflamable; others there are which grow hard, and are not to be kindled into a Flame. They are easily to be dissolved in Fountain-water, (the Gum of Rhubarb and the Leaves, for example) and do sparkle when put into a Flame: Which two Natures argue a serous or watrish Part in them. Again, put into a Flame, they melt and become as it were liquid, and ductile; which shews the caseous part in them. And because they will not flame, it is an Argument of their Leanness, and Scarcity of Oil. All three put together plainly evince Gums to be coagulate Juices.

In August I have observed the Clusters, both green and ripe, of *Periclymenum Ger.* very leaky; which upon nearer and heedful Inspection I found to be a thin clammy Juice, or liquid Gum, which fell down upon the Leaves, and keeps its liquid Form there.

Here the purple Juice seems to be a Whey separated from the liquid Gum: But I am of opinion it is a distinct Liquor.

Again the red Threads of *Rorella* end, or are topped, with little Bags; which being compressed do yield a purple Juice (as we above noted in the *Hypericum*) and those small Buttons on the very tops of those Threads, are encompassed with small transparent Pearls or Drops of a liquid Gum. They abide in this Form the hottest Summer's Day like Dew, whence also the Plant has its Name; and upon the least Touch cleave to your Fingers, and draw out into long Threads like Bird-lime.

In like manner a liquid Gum (but that it stands not upon so long Threads, and is much thicker bedewed) you may observe upon *Pinguicula*. *Note well*, That the small Drops and Threads, or Hairs, in either of these two Plants, are to be seen upon the uppermost or inmost side of the Leaf; and the utmost and undermost is smooth or void of them; which is something contrary to all other Plants I have observed.

Methought I observed about the middle of August, the Chats of the Alder to be gummy. Perhaps it did exsudate from the Plant itself; as I guess the Honey-Fall, or gummy Dew to be observed upon the Leaves of the Oak, &c. are nothing else.

The American, or Indian Rhubarb, sown in our Gardens is the only Plant that I have met with, or ever saw, which yielded a Gum; and yet because it is of the very kind with our common Sorrels and *Lapatha*, I believe it not impossible, that even from our own Store, Herb-Gums might someways or other be had.

I say, that off the Stalk, or indeed off the Leaves of the Indian Rhubarb, I have gathered an Ounce at a time in June, of very white, clear, and hard Gum: In both those Years I observed it to flower with us, as 1670, and in that Year it did not, 1669. It exsudates from all Parts of the Stalk and Ribs, on (*Note well*) the underside of the Leaf itself. I gathered some in form of good big Drops, others as though the Stalk had been besmeared with it, others shot into long and twisted Wires, or Icicles. Moreover I observed, that the cankered Orifices or Places where the Gum had burst forth, might be followed into the Stalk with a Knife, and that through the Skin, in certain Places, I could see that the Juice within the Plant was turned gummy, and looked clear like Ice.

It is the Experiment of Mr Fisher, that the clear and desecated Juices of most Plants have more or less Redness in them. Again, that the dryed Root of Acetosa (a Plant of the Family with Rhubarb, which may well be called the Indian Sorrel, or sour Docken) boiled, doth dye Water with a fair red Colour. And I have observed, that the unripe Seeds of Rhubarb yield a very fair and deep Purple, I mean the Husk of them. Consider what hath been said above of *Rorella*, and the *Hypericum* kind, concerning their Purple Juices yielding Blebs. *Note*, also here to this Purpose what we have set down above, Rhubarb, Sorrel, &c. do when they decay turn red.

The Juice extracted from the Roots of our English Rhubarb, by a Tincture of fair Water steam'd away, is nothing else but a lean uninflammable Gum; and tho' it differ in Colour, perhaps from the yet woody Parts in it, as being of a deep Liver Colour, from the exsudating Gum, yet in other Natures, as this of being uninflamable, ductile in the Flame of a Candle, &c. it agrees with it. I may not omit, that the repeated Cuts I gave the Stalk, on purpose to have the Gum that way, failed my Expectation. This Gum is sweet, or rather of no Taste at all.

To this purpose I remember in Summer time to have seen, even the Juice of Apples spontaneously gellied in *Languedoc*, and the Apples to look clear and hard like Ice, whence they call that sort of Apple, *Pome Gellee*, or the frozen Apple: Tho' indeed, it is nothing else but the breaking or coagulating of the Juice in some Spots of it; for it is rare to see one of them all over so.

We may here give a probable Reason why a gentle Infusion or Maceration of Rhubarb is a very sure Purge, but the Substance or Powder of Rhubarb, or a Decoction thereof, will have a quite contrary Effect, and bind. We may, I say, think that the sharp and tart Juice in Rhubarb, wherein its purging Faculty lies, is by a gentle Infusion so extracted, that it turns not to Gum in our Stomach; for I cannot think that the sour Juice of Rhubarb is a specifically distinct Liquor from the Gum, which I believe to be only an accidental Coagulation.

Green Plums, or Sloes, do often break forth with a Gum, which is clear and transparent; and it seems to hasten, if not ripen, at least the red Colour. I have cut them, to the end that I might have gathered Gum in the Wounds; which indeed I did, but yet long after, when the Wounds seemed to be cankered, and that but in a small Quantity to what they voluntarily spend.

Lauro-cerasus, a beautiful Winter-green, which we have adopted to adorn our Court-Walls with, yields a clear Gum very plentifully; it is very white and very clear.

There are other sorts of Juices, which will not of themselves, that I have observed, exsudate out of the Wounds of their respective Plants.

I wrenched and wounded the Holly the latter End of March, and yet after some Days of warm and open Weather, I could not perceive the least stirring of Juice: the latter End of May the Bark begins to be full of Lime, which you may try by pressing a Piece of it betwixt your Fingers, and when you would take them off, the Juice or Lime draws out into Hairs, and follows your Fingers, cleaving to them like small Threads.

This Lime or Juice is separated or taken out of the Bark

thus: Peel off the Bark the Months of May, June or July; for then it comes easily away, and most abounds with Juice: Boil the Bark in fair Water, until it be so tender, that the utmost thin-grey Bark or Membrane, peel easily off; lay it so peeled, and cover it over with green Nettles or Fern, or such like, *S.S.S.*, in a Cellar for about ten Days, where it will ferment or rot, and become mouldy: Take them out and beat them well in a Mortar to a Paste, and roll them up into small Hand-balls, and in a running Spring wash them clean from all the woody or sticky Parts; which is effected by pulling and teasing them. But *Note well*, that great Care is to be taken in the Washing of the Balls; for besides that they must, if possible, be forthwith washed, the Lime will all get from you, except you so order the Matter, by engaging it with your Fingers, that it entangle. You would imagine, that upon breaking one of the Balls, that there was little or no Lime in them, so freely they moulder and crumble. After they once engage thoroughly, it will endure washing; and the clearer you take away the woody Parts, the better it is.

In cutting the tender Tops of Elder, the latter End of May, there will a stringy Juice follow your Knife, and draw out in Threads, somewhat like Bird-lime, or the Juice of Holly: It seems to be in certain Veins just within the Circle of Teeth or Wood.

Further, the dissected Veins of many Plants afford us Oil, that is, such a Juice, which being rubbed betwixt one's Fingers, is not at all clammy, but makes them greasy and glib. Some of it stiffens not, as far as I have yet experienced; yet I believe it to be coagulate and mixed.

We will instance in the Juice of *Helenium, sive Enula Campana J.B.* You may take it off with a clean Knife, whereon it looks like Oil mixed with Water; that is, the thin or dilute Juice of the Plant, springing up out of the Wound, together with the Oil. The like Experiment may be made upon *Cicuta.* The Juice of *Angelica sativa Park.* I found altered after a Year's keeping, and grown very Limy.

Tapsus barbatus Ger. If you strip off the Leaves in June, it seems to yield an oily Juice, but very much thinned with the watry one. It springs freely enough; it is of a dark green Colour, and I took it in Wheat-flower, and made it up in Cakes.

Also the Fruits of many Plants afford Oil, as *Oliva*, *Baccæ Lauri*, *Hederæ*, *Juniperi*, *Cornus Fœminæ* &c.

The Pulp of most Seeds seem to abound with this oily Juice, and at sometime before their Maturity it is liquid and visible in them, in the form of a Milk.

Helleborus niger syl. adulterinus, etiam Hyeme virens J.B. The Seeds of this Plant, the latter End of May, are very milky, and by Insolation are easily formed into Cakes, which are yet very oily, and being long kept, I have exposed to the Flame of a Candle, which they received and burnt freely; sparkling not very much, and not then neither being clammy at all. One thing I must not omit, that this Milk or Juice of the Seeds is of a very fiery and stinging Nature; for where I cut the Seeds out of the green Pods, they struck my Eyes no otherwise than Onion is wont to do. Moreover, the Tops of my Fingers, which were wetted with this Juice, did boaken and ach, as when after extreme Cold one has the Hot-Ach in them; and that Pain continued in them for several Days; at length the Skin of my Finger's End peeled off.

Diacodium Album is a Medicament of the Seeds of Poppy.

There are other oily Juices, which after Coagulation harden and are called Rosin; and such our Ivy yields abundantly. Hither also may be referred the Juice of *Juniperus vulgaris, baccis parvis purpureis J.B.* which is a hard fat Juice, and not much gummy.

In the Chops of Ivy made in March, there did exsudate a thick Matter like Barm, yellowish and greasy; It melted like Oil betwixt my Fingers, not having the least Clamminess then perceivable. In process of time it hardened and crusted on the Wounds like coarse brown Sugar. It burns with a lasting Flame, and smells very strong.

Also on the topmost Leaves of *Lactuca syl. Costa spinosa C.B.* In July many small Drops or Pearls of an oily Juice, like coagulated and hardened Rosin, are plain to be discerned, especially with a single Microscope: They are of an Amber Colour, and transparent; easily to be wiped off, as being an oily Juice exsudated. And I am of the Mind, that even the blue flower of ripe Plums is nothing else but a fine resinous Coagulation of the transudated Juice.

On the under-side of the Leaves, and all over the Stalk of *Bonus Henricus J.B.* do stick infinite small transparent Pearls: Those clear Drops are hard to the Touch, and feel like greasy Sand; not clammy, and therefore it was well called unctuous by C.B. and we put this spontaneously exsudated Juice amongst the resinous Coagulations.

The Juices of Plants are also varied and distinguished by Fermentation. And not only the Juices of Fruits are to be wrought, or set a working; as of the Apple, Pear, Briar, Grape, &c. as is well known: But there is an artificial Change, *viz.*, Malting, to be made even in the Seeds of Plants, so as to make them spend freely, or let go their Juices, and communicate them to common Water, and receive a Ferment: Also the Juice of the Roots of *Glicyrrhiz* will ferment: Also the Juice of Cane, as Sugar. Again, the tapped Juices of Vegetables (wherein my Observations are limited) are susceptible of a Ferment. As for Instance:

The 21st of April 1665, about eight in the Morning, I bored a Hole in the Body of a fair and large Birch, and put in a Cork, with a Quill in the middle; After a Moment or two it began to drop, but yet very softly: Some three Hours after, I returned, and it had filled a Pint Glass, and then it dropped exceeding fast, *viz.*, every Pulse a Drop. This Liquor is not unpleasant to the Taste, and not thick or troubled: Yet it looks as though some few Drops of Milk were spilt in a Bason of Fountain-Water. There are many Ways of fermenting or setting this Juice a working, that is, of keeping it from coagulating.

The Maple, both that which is miscalled the Sycamore, and the lesser, bleed a fermentable Juice copiously, in the breaking up of hard Frosts.

Also the Willow, Walnut, Poplar, Wicking, are all said to bleed in their Seasons a vinous Juice.

To extract the Juice of Vegetables, as *Opium*, for Example, (as is usual in the best Preparations and Methods of making Laudanum) with Spirit of Wine, is not probably to separate any one part of that coagulate Juice from the other, as the Serum or Whey, for example, from the caseous part of the Juice, but only to depurate or desecate the *Opium*: For *S.V.* says Mr Boyle, will dissolve *Gum. Lac. Benzoin*, and the resinous Parts of *Jallap*, and even of *Guiacum*, which are Cogulations and mixed Juices; and the same we may think of the Juices that are extracted by *S.V.* from other Herbs that are mixed.

Also those other Ways of roasting and drying Juices upon Plates over a gentle Fire, until they will rub to Powder, gives no great Satisfaction to me, that the *Narcosis* of *Opium*, for example, is gone or separated, because the dried Juice less offends the Nose; that is, smells not so strong.

The Whey of *Lact. syl.* will be only dissolved in cold Water, the Curds wholly refusing to mix with it: So that simple Water perhaps is the best Menstruum, and really separates what *S.V.* only depurates.

MARTIN LISTER, JOHN RAY, AND SIR TANCRED ROBINSON: from the *Philosophical Transactions of the Royal Society* (*c.* 1693)

Fungi
By Martin Lister

As I passed through Marton Woods, under Pinno Moor in Craven, Aug. 18, 1672, I found an infinite number of Mushrooms, some withered, and others new sprung and flourishing.

from Observation of the present state of the Earth, and of the site and condition of the Marine Bodies which are lodged within and upon it, shewn that they could not possibly be reposed in that manner by particular Inundations: by the Seas receding and shifting from place to place: nor by any of the other means there proposed: I pass next on to search out the true means; and to discover the Agent that did actually bring them forth, and disposed them into the Method and Order wherein we now find them. To which purpose I have recourse again to the Observations; for by their Assistance this Matter may be rightly and fully adjusted. So that I shall only proceed, as hitherto, to make Inferences from them; which Inferences, in this Part are all *affirmative*. Of these, the first is,

That these Marine Bodies were born forth of the Sea by the Universal Deluge: and that, upon the return of the Water back again from off the Earth, they were left behind at Land.

This is a Proposition of some weight and consequence; upon which Account I shall be somewhat prolix and particular in the Establishment of it: careful and exact in conferring every circumstance of these Marine Bodies, to see how they square with it: and shall not dismiss it till I have evinced that those which I prest, in the precedent Part, as Objections against the several ways there propounded, all fall in here, and are the clearest and most convincing Arguments of the truth hereof; that this, and this alone, does naturally and easily account for all those Circumstances: And fairly takes off all Difficulties. Which Difficulties I propose at large, and particularly those which have of late been urged, by some Learned Men, as proofs that these Bodies were not left behind by the Deluge, shewing of how little Validity they are.

Which being dispatch'd, I return back to my Observations, and proceed upon them to represent the Effects that the Deluge had upon the Earth, and the *Alterations* that it wrought in the Globe; some whereof were indeed extraordinary. *These* I distribute into two Classes; the *first* of which will contain *those* that are *only* probable, and of which we have some reasonable Intimations, but not an absolute and demonstrative Certainty, the Proofs whereon they depend being more remote. And these I shall wholly wave at present, and not crowd this shorter Treatise with the Relation of them, reserving that room for *those* of the *second* Class, which are those whereof we have a plain and undeniable Certainty: those which flow directly and immediately from the Observations, and which are so evident, that 'tis impossible these Marine Bodies could have been any ways lodged in such manner, and to so great depths, in the Beds of Stone, Marble, Chalk, and the rest, had not these Alterations all happened. Namely

That during the time of the Deluge, whilst the Water was out upon, and covered the Terrestrial Globe, All the Stone and Marble of the Antediluvian Earth: all the Metalls of it: all Mineral Concretions; and, in a word, all Fossils what ever that had before obtained any Solidarity, were totally dissolved and their constituent Corpuscles all disjoyned, their Cohæsion perfectly ceasing. That the said Corpuscles of these solid Fossils, together with the Corpuscles of those which were not before solid, such as Sand, Earth, and the like: as also all Animal Bodies, and parts of Animals, Bones, Teeth, Shells: Vegetables, and parts of Vegetables, Trees, Shrubs, Herbs: and, to be short, all Bodies whatsoever that were either upon the Earth, or that constituted the Mass of it, if not quite down to the Abyss, yet at least to the greatest depth we ever dig: I say all these were assumed up promiscuously into the Water, and sustained in it, in such manner that the Water, and Bodies in it, together made up one common confused Mass.

That at length all the Mass that was thus borne up in the Water, was again precipitated and subsided towards the bottom. That this Subsidence happened generally, and as near as possibly could be expected in so great a Confusion, according to the Laws of Gravity: *that* Matter, Body or Bodies, which had the greatest quantity or degree of Gravity, subsiding first in order, and falling lowest: *that* which had the next, or a still lesser degree of Gravity, subsiding next after, and settling upon the precedent: and so on in their several Courses; *that* which had the least Gravity sinking not down till last of all, settling at the Surface of the Sediment, and covering all the rest. That the Matter, subsiding thus, formed the *Strata* of Stone, of Marble, of Cole, of Earth, and the rest; of which Strata, lying one upon another, the Terrestrial Globe, or at least as much of it as is ever displayed to view, doth mainly consist. That the Strata being arranged in this order merely by the disparity of the Matter, of which each consisted, as to Gravity, that Matter which was heaviest descending first, and all that had the same degree of Gravity subsiding at the same time: and there being Bodies of quite different Kinds, Natures, and Constitutions, that are nearly of the same specifick Gravity, it thence happened that Bodies of quite different kinds subsided at the same instant, fell together into and composed the same *Stratum*. That for this reason the Shells of those Cockles, Escalops, Perewinkles, and the rest, which have a greater degree of Gravity, were enclosed and lodged in the *Strata* of Stone, Marble, and the heavier kinds of Terrestrial Matter: the lighter Shells not sinking down till afterwards, and so falling amongst the lighter Matter, such as Chalk, and the like, in all such parts of the Mass where there happened to be any considerable quantity of Chalk, or other Matter lighter than Stone; but where there was *none*, the said Shells fell upon or near unto the Surface: and that accordingly we now find the lighter kinds of Shells, such as those of the Echine, and the like, very plentifully in Chalk, but of the heavier kinds scarcely one ever appears, these subsiding sooner, and so settling deeper, and beneath the *Strata* of Chalk.

That Humane Bodies, the Bodies of Quadrupeds, and other Land-Animals, of Birds, of Fishes, both of the Cartilaginous, Squamose, and Crustaceous kinds; the Bones, Teeth, Horns, and other parts of Beasts, and of Fishes: the Shells of Land-

Snails: and the Shells of those River and Sea Shell-Fish that were lighter than Chalk &c. Trees, Shrubs, and all other Vegetables, and the Seeds of them: and that peculiar Terrestrial Matter where-of these consist, and out of which they are all formed: I say all these (except some Mineral or Metallick Matter happened to have been affix'd to any of them, whilst they were sustained together in the Water, so as to augment the weight of them) being, bulk for bulk, lighter than Sand, Marl, Chalk, or the other ordinary Matter of the Globe, were not precipitated till the last, and so lay above all the former, constituting the Supreme or outmost Stratum of the Globe.

Sir Richard Buckley (or Bulkeley), Samuel Foley, and Thomas Molyneux:

from the *Philosophical Transactions of the Royal Society* (*c.* 1693)

The Giant's Causeway in Ireland
By Sir Richard Buckley (or Bulkeley)

This Description of the Giant's Causway, I received from a Scholar and a Traveller, who went on purpose the last summer 1692, with the Bishop of Derry to see it. It is in the County of Antrim, about 7 Miles East of Colrain, and 31 Miles to the East of the Mouth of the River Derry. The Coast there is a very great Height from the Sea: And from the Foot of the Precipice, there runs out Northward into the main Ocean, a raised Causway of about 80 Foot broad, and about 20 Foot high above the rest of the Strand; its Sides are perpendicular, it was about 200 Foot in view to the Sea-Water.

This whole Causway consists all of Pillars of Perpendicular Cylinders, Hexagons and Pentagons, of about 18 and 20 Inches Diameter, but so justly shot one by another, that not any thing thicker than a Knife will enter between the sides of the Pillars. When one walks upon the Sand below it, the side of this Causway has its Face all in Angles, the several Cylinders (pardon the Impropriety of the Word) having some two, some three of their Sides open to View. The very vast high Precipice does also consist of Cylinders: tho' some shorter and some longer: And all the Stones that one sees on that coast, whether single or in Clusters, or that rise up any where out of the Sand, are all Cylinders, tho' of ever so different Angles; for there are also Four-Squared upon the same Shore.

By Samuel Foley

The Giant's Causway is somewhat more than 8 English Miles North-East from the Town of Colrain, and about 3 from the Bush Mills, almost directly North. It runs from the bottom of an high Hill into the Sea, no Man can tell how far, but at Low Water the Length of it is about 600 Foot, and the Breadth of it, in the broadest Place 240 Foot, in the narrowest 120 Foot. It is very unequal likewise in the Height, in some Places it is about 36 Foot high from the Level of the Strand, and in other Places about 15 Foot.

It consists of many thousand Pillars, which stand most of them perpendicular to the Plain of the Horizon close to one another, but we could not discern whether they do run down under ground like a Quarry or no. Some of them are very long and higher than the rest, others short and broke; some for a pretty large Space of an equal Height, so that their Tops make an even plain Surface, many of them imperfect, crack'd and irregular, others entire, uniform and handsome, and these of different Shapes and Sizes. We found them almost all Pentagonal or Hexagonal only we observed that a few had 7 sides and many more Pentagons than Hexagons, but they were all irregular: For none that we could observe had their sides of equal Breadth; the Pillars are some of them 15, some 18 Inches, some 2 Foot in Diameter, none of them are one entire Stone, but every Pillar consists of several Joints or Pieces, as we may call them, of which some are 6, some 12, some 18 Inches, some 2 Foot deep.

These Pieces lie as close upon one another as 'tis possible for one Stone to lie upon another; not jointing with flat Surfaces, for when you force one off the other, one of them is always Concave in the middle, the other Convex. There are many of these kind of Joints, which lie loose upon some part of the Causway, and on the Strand, which were blown or washed off the Pillars. These Joints are not always placed alike, for in some Pillars the Convexity is always upwards, and in others it stands always downwards. When you force them asunder, both the Concave and the Convex Superficies are very smooth, as are also the sides of the Pillars which touch one another, being of a whitish Free-stone Colour, but a finer closer Grit; whereas when we broke some Pieces off them, the Inside appeared like dark Marble.

The Pillars stand very close to one another, and tho' some have 5 Sides, and others of them 6, yet the Contexture of them are so adapted, that there is no Vacuity between them; the Inequality of the Numbers of the sides of the Pillars, being often in a very surprizing and a wonderful Manner, throughout the whole Causway, compensated by the Inequality of the Breadths and Angles of those Sides; so that the whole at a little Distance looks very Regular, and every single Pillar does retain its own Thickness, and Angles and Sides, from Top to Bottom.

Those Pillars which seem to be entire as they were originally, are at the Top flat and rough, without any Graving or Striate Lines; those which lie low to the Sea, are washed smooth; and others that seem to have their natural Tops blown or washed off, are some concave, and others convex.

The high Bank hanging over the Causway on that side which lies next it, and towards the Sea, seems to be for the most part composed of the common sort of Craggy Rock, only we saw a few irregular Pillars on the East-Side, and some farther on the North, which they call the *Looms*, or *Organs*, standing on the side of a Hill; the Pillars in the middle being the longest, and those on each side of them still shorter and shorter: But just over the Causway, we saw as it were the Tops of some Pillars appearing out of the sides of the Hill, not standing, nor lying flat, but sloping.

We suppose each Pillar, throughout the Causway, to continue the same to the very bottom, because all that we saw on the Sides were so.

N.B. The several sides of one and the same Pillar are as in the Planes of Chrystals, of very unequal breadths or lengths, call it either, when you measure them Horizontally; and that in such as are Hexagonal a broader Side always subtends, or is opposite to, a narrower, which sort of Geometry Nature likewise observes in the Formation of Chrystals.

By Thomas Molyneux

Among the several Figur'd Stones already described by Authors, I find none that has more Agreement with those which compose our Giant's Causway than the *Entrochos*, the *Astroites* or *Lapis Stellaris*, and the *Lapis Basanus* or *Basaltes*: And yet for all the great Resemblance they have in some Particulars, they differ very much in others.

The *Entrochos* agrees with the Pillars of our Causway in that it is a stony Substance, formed by Nature Column-wise, and consisting of 20 or 30 several *Internodia*, or Joints, set one a-top of another; but then it differs in that its outward Shape is round and Cylindrical; in its having a Hole, or Pith, run from top to bottom through all the Joints; in the setting on, or way of fitting one Joint to another; and in its Size and Magnitude.

The *Astroites* or *Lapis Stellaris* is not only shaped Column-wise, as the *Entrochos*, and jointed with several *Internodia* closely adjusted to one another, but its Sides are Angular. But then it must be observed that the Sides of the *Astroites* are always sulcated or a little furrowed, and are constantly Pentagons; whereas the Irish-Stone has its Sides perfectly smooth, and plane, and sometimes in Hexagons and Heptagons as well as

Pentagons. Moreover the *Astroites* has Furrow'd and Portuberant Rays striking from its Center, somewhat as they draw a Star, whence it has its Name; that adapting the Concavites and Convexites together, cause the Cohesion of the Joints to one another; whereas the internal Superficies of the *Internodia* in our Irish-Stone sends forth no sort of Rays from its Center, and unite to one another by a quite different Articulation. For besides what Dr Foley remarks of the bottom or top of each Joint, having a large round Concavity or Convexity that extends it self from the Center of the Stone within an Inch or two of the Angular Circumference; examining two Joints that were sent up from the Place hither to Dublin, I observed likewise, that the bottom or top of each Joint round this Concavity or Convexity either rises with an eminent Verge, or Ridge, if it be Concave in the Middle, or if it be Convex, is hollowed with such a sort of Groove, as to receive closely into it all the eminent Ridge of the next Joint either above or below it: so that each Superficies in the Articulations adapt themselves on all sides so exactly one to the other, as 'tis possible for two Bodies, that are only Contiguous and not Cohering.

The *Astroites* also, as well as the *Entrochos*, differs extremely from our Stone in its Size, or Magnitude; for the largest that is found of either of those kinds, do not much exceed the thickness of a Man's Thumb, whereas our Columns are some of them two Foot in Diameter. Yet this disproportion of Bulk is not so considerable a Difference, since we observe that Nature affects the like Disparity in other of her Works, and those too nearly allied, and evidently of the same Tribe, or Family. Our small Jointed Rushes or Reeds, and the largest East Indian Bambou, one of which I remember to have seen in Holland above 26 Foot high, and as thick as a Man's Middle, are yet Plants of the same Species and Class.

But nothing among all the Fossil Tribe that I have seen or read of, comes so nigh in all respects in its Formation, Substance, Size, Way of growth or Manner of standing, &c. to the Columns whereof the Causway is composed, as the *Lapis Basaltes Misneus*, described by Kentmannus in Gesner *de Figuris Lapidum*, whereof he says there is a great large Bed within three Miles of Dresden in Saxony. He gives the following Account of it thus in his own Words—*Lapides Angulosi plures coagmentati Basalton repræsentat, qui crescet forma & magnitudine Fici mediocris, Singularis quidem sed Copiosus atque ita Junctus Coaptatusque, veluti ab Arculario arte commissus esset; septem, sex, quinque, nonnunquam sed rarius quatuor Angulorum: Omnino Figura Trabis erectæ, foris Levis, & Tactu minime Asper, Ferrugineus, Ponderosus, Duritie velut Adamantis; Hi Lapides sic Coagmentati e terra Ulnas decem & septem extant; quanto spatio intra Terram condantur, nemini adhuc exploratum est.* But, I find this Difference between these and the Misnean *Basaltes*, that its Columns were one entire Piece from top to bottom, whereas our Irish *Basaltes*, is composed of Columns divided into many Joints. So that I think it may not improperly be called, to distinguish it from this and all other Fossils; *Lapis Basaltes vel Basanos Maximus Hibernicus Angulis minimum Tribus, plurimum Octo constans; crebris Articulis sibi invicem affabre conjunctis, sed facile separabilibus, Geniculatus.*

Whether our Irish *Basaltes* can pretend to the Name *Basanos*, on the same account the Misnean does, from the Greek Word *Basanizo, Exploro*, because it has the Property of the Touch Stone, that shews by Lines drawn with Metals on its smooth Surface, which are Genuine, and which Adulterate, I cannot positively say; because those Pieces I have, are so rough, that unless some part of the Superficies were artificially polished, the Experiment cannot be made: Yet I have reason to believe it would succeed, were the Stone polished; because I find Black Marble in general, so it be of a close Texture and hard, as this is, partakes of that Property.

ASTRONOMY

John Wallis:
from the *Philosophical Transactions of the Royal Society* (1693)

To find the Parallax of the Fixed Stars

Give me leave to suggest a Speculation, which hath been in my Thoughts these forty Years or more; but I have not had the Opportunity of reducing it to Practice: It is concerning the Parallax of the fixed Stars, as to the Earth's annual Orbit.

Galileo complains of it a great while since (in his *Systema Cosmicum*) as a thing not attempted to be observed with such Diligence as he could wish; and I doubt we have the same Cause of complaining still. I know that Dr Hook and Mr Flamsteed have attempted somewhat that way, but have desisted before they came to any thing of Certainty. What hath been done to that purpose Abroad I know not.

Galileo hath suggested divers Things considerable in order to it; as, the Times of Observation, the Stars to be observed, and the Manner of observing them; which yet I doubt is not practicable. That which occurred to my Thoughts upon these Considerations, was to this purpose: That some circumpolar Stars (nearer to the Pole of the Equator than is our Zenith, and not far from the Pole of the Zodiack) should be made choice of for this purpose. And in case the meridional Altitude be discernibly different at different Times, so will also be their utmost East and West Azimuth, which may be better observed than their Rising or Setting: And this will not be obnoxious to the Refraction, as is the meridional Altitude (for though the Refraction do affect the Altitude, yet not the Azimuth at all;)

and we may here have choice of Stars for the purpose; which, in Observations from the Bottom of a Well, we cannot have; being there confined to those only which pass very near our Zenith, tho' very small Stars.

I would then take for granted, as a thing at least very probable, that the fixed Stars are not all (as was wont to be supposed) at the same Distance from us, but the Distance of some vastly greater than of others; and consequently, though as to the more remote, the Parallax may be undiscernible, it may, perhaps, be discernible in those that are nearer to us.

And those we may reasonably guess (though we are not sure of it) to be nearest to us, which to us do appear biggest and brightest, as are those of the First and Second Magnitude, and there are at least of the Second Magnitude pretty many not far from the Pole of the Ecliptick (as that in particular in the Shoulder of the Lesser Bear): And, in case we fail in one, we may try again and again on some other; which may chance to be nearer to us than what we try first. And Stars of this Bigness may be discerned by a moderate Telescope, even in the Day-time; especially when we know just where to look for them.

The Manner of Observation, I conceive may be thus. Having first pitched upon the Star we mean to observe, and having then considered (which is not hard to do) where such Star is to be seen in its greatest East or West Azimuth; it may be then convenient to fix (very firm and steadily on some Tower, Steeple, or other high Edifice, in a convenient Situation) a good Telescopick Object-Glass in such Position as may be proper for viewing that Star. And at a due Distance from it, near the Ground, build on purpose (if already there be not any) some little Stone Wall, or like Place, in which to fix the Eye-Glass, so as to answer that Object-Glass: And having so adjusted it, as through both to see that Star in its desired Station (which may best be done while the Star is to be seen by Night in such Situation, near the Time of one of the Solstices) let it be there fixed so firmly, as not to be disturbed (and the

Place so secured, as that none come to disorder it) and care be taken so to defend both the Glasses, as not to be endangered by Wind and Weather. In which Contrivance, I am beholden to Mr John Caswel, M.A. of Hart-Hall in Oxford, for his Advice and Assistance, with whom I have many Years since communicated the whole Matter.

This Glass being once fixed (and a Micrometer fitted to it, so as to have its Threads perpendicular to the Horizon, to avoid any Inconvenience which might arise from Diversity of Refraction, if any be) the Star may then be viewed from Time to Time (for the following Year, or longer) to see if any Change of Azimuth can be observed.

This I thought fit to recommend to your Consideration, who do so well understand Telescopes, and the Managery of them: But when I suggest (as a convenient Star for this purpose) the Shoulder of the Lesser Bear (as being the nearest to the Pole of the Zodiack of any Star that is of the First or Second Magnitude) I do not confine you to that Star; but (without retracting that) suggest another; namely, the Middle Star, in the Tail of the Great Bear, which (though somewhat further from the Pole of the Zodiack) is a brighter Star than the other, and may be nearer to us.

But I do it principally upon this Consideration; namely, That there is adhering to it a very small Star, (which the Arabs call *Alcor*, of which they have a proverbial Saying, when they would describe a sharp-sighted Man, That he can discern the Rider on the Middle Horse of the Wayn: And of one who pretends to see small Things, but overlooks much greater, *Vidit Alcor at non Lunam Plenam*:) Which Hevelius in his Observations finds to be distant from it about nine Minutes, and five or ten Seconds: So that besides the Advantage of discovering the Parallax of the greater Star, if discernible, the Difference of the Parallax of that and of the lesser Star (being both within the Reach of a Micrometer) may do our Work as well. For if that of the greater Star be discernible, but that of the lesser be either not discernible or less discernible, their different Distances from each other at different Times of the Year may perhaps (without farther Apparatus) be discerned by a good telescope of a competent Length, furnished with a Micrometer, if carefully preserved from being disordered in the Intervals of the Observations; and discover at once, both that there is a Parallex, and that the Fixed Stars are at different Distances from us; wherein, that I may not be mistaken, my Meaning is not that the Instrument or Micrometer should be removed for the observing of the Lesser Star, but that (when the Azimuth of the Greater Star is taken) by a Micrometer (consisting of divers fine Threads parallel and transverse) may (at the same time) be observed the Distance of the two Stars, each from other, in that Position (both being at once within the Reach of the Micrometer) Which Distance (the Instrument remaining unmoved) if it be found (at different Times of the Year) not to be the same, this will prove that there is a different Parallex of these two Stars.

This latter Part of the Observation (of their different Distances at different Times) I suggest as more easily practicable, though not so nice as the former: For it may be done, I think, without any further Apparatus there than a good Telescope of ordinary Form, furnished with a Micrometer, (this being carefully kept unvaried during the Interval of these Observations.) And if this Part only of the Observation (without the other) be pursued, it matters not though the two Observations (near the two Solstices) be, one at the Eastern, the other at the Western, Azimuth (whereby both may be taken in the Night-time) for the Distance must (at both Azimuths) be the same. If, after observing the Azimuth of the greater Star, it be necessary to move the Micrometer for measuring its Distance from *Alcor*, that may be done another Night, (and it is not necessary to be done at one Observation) for that Distance cannot be discernibly varied in a Night or two.

METEOROLOGY

Robert Hooke:

from Thomas Sprat's *History of the Royal Society* (1667)

A Method for Making a History of the Weather

For the better making a History of the Weather, I conceive it requisite to observe,

1. The Strength and Quarter of the Winds, and to register the Changes as often as they happen; both which may be very conveniently shewn, by a small Addition to an ordinary Weather-cock.

2. The Degrees of Heat and Cold in the Air; which will be best observed by a sealed *Thermometer*, graduated according to the Degrees of *Expansion*, which bear a known Proportion to the whole Bulk of Liquor, the beginning of which Gradation, should be that Dimension which the Liquor hath, when encompassed with Water, just beginning to freeze, and the Degrees of Expansion, either greater or less, should be set or marked above it, or below it.

3. The Degrees of Dryness and Moisture in the Air; which may be most conveniently observed by a Hygroscrope, made with the single Beard of a wild Oat perfectly ripe, set upright and headed with an Index, after the Way described by Emanuel Magnan; the Conversions and Degrees of which may be measured by Divisions made on the rim of a Circle, in the Center of which, the Index is turned round: The Beginning or Standard of which Degree of Rotation, should be that, to which the Index points, when the Beard, being throughly wet, or covered with Water, is quite unwreathed, and becomes strait. But

because of the Smalness of this Part of the Oat, the Cod of a wild Vetch may be used instead of it, which will be a much larger Index, and will be altogether as sensible of the Changes of the Air.

4. The Degrees of Pressure in the Air; which may be several Ways observed, but best of all with an Instrument with Quicksilver, contrived so, as either by means of Water, or an Index, it may sensibly exhibit the minute Variations of that Action.

5. The Constitution and Face of the Sky or Heavens; and this is best done by the Eye; here should be observed, whether the Sky be clear or clouded; and if clouded, after what Manner; whether with high Exhalations or great white Clouds, or dark thick ones. Whether those Clouds afford Fogs or Mists, or Sleet, or Rain, or Snow, &c. Whether the under side of those Clouds be flat or waved and irregular, as I have often seen before Thunder. Which way they drive, whether all one Way, or some one way, some another, and whether any of these be the same with the Wind that Blows below; the Colour and Face of the Sky at the rising and setting of the Sun and Moon; what Haloes or Rings may happen to encompass those Luminaries, their Bigness, Form and Number.

6. What Effects are produc'd upon other Bodies: As what Aches and Distempers in the Bodies of Men; what Diseases are most rife, as Colds, Fevers, Agues, &c. What Putrefactions or other Changes are produc'd in other Bodies; as the sweating of Marble, the burning blue of a Candle, the blasting of Trees and Corn; the unusual Sprouting, Growth, or Decay of any Plants or Vegetables; the Putrefaction of Bodies not usual; the Plenty or Scarcity of Insects; of several Fruits, Grains, Flowers, Roots, Cattel, Fishes, Birds, any thing notable of that Kind. What Conveniences or Inconveniences may happen in the Year, in any kind, as by Floods, Droughts, violent Showers, &c. What Nights produce Dews and Hoar-Frosts, and what not?

7. What Thunders and Lightnings happen, and what

Effects they produce; as souring Beer or Ale, turning Milk, killing Silkworms, &c?

Any thing extraordinary in the Tides; as double Tides, later or earlier, greater or less Tides than ordinary, rising or drying of Springs; Comets or unusual Apparitions, new Stars, *Ignes fatui* or shining Exhalations, or the like.

These should all or most of them be diligently observed and registered by some one, that is always conversant in or near the same Place.

Now that these, and some others, hereafter to be mentioned, may be registered so as to be most convenient for the making of Comparisons, requisite for the raising Axioms, whereby the Cause or Laws of Weather may be found out; it will be desirable to order them so, that the Scheme of a whole Month may at one View be presented to the Eye: And this may conveniently be done on the Pages of a Book in Folio, allowing fifteen Days for one side, and fifteen for the other. Let each of those Pages be divided into nine Columns, and distinguished by perpendicular Lines; let each of the first six Columns be half an Inch wide, and the three last equally share the remaining of the Side.

Let each Column have the Title of what it is to contain in the first at least, written at the Top of it: As, let the first Column towards the left hand, contain the Days of the Month, or Place of the Sun, and the remarkable Hours of each Day. The second, the Place, Latitude, Distance, Ages and Faces of the Moon. The third, the Quarters and Strength of Winds. The fourth, the Heat and Cold of the Season. The fifth, the Dryness and Moisture of it. The sixth, the Degrees of Pressure. The seventh, the Faces and Appearances of the Sky. The eighth, the Effects of the Weather upon other Bodies, Thunders, Lightnings, or any thing extraordinary. The ninth, general Deductions, Corollaries or Syllogisms, arising from the comparing of several Phænòmena together.

That the Columns may be large enough to contain what they

are designed for, it will be necessary, that the Particulars be expressed with some Characters, as brief and compendious as is possible. The two first by the Figures and Characters of the Signs commonly used in Almanacks. The Winds may be exprest by the Letters, by which they are exprest in small Sea-Cards; and the Degrees of Strength, by 1, 2, 3, 4, &c. according as they are marked in the Contrivance in the Weather-cock. The Degrees of Heat and Cold may be exprest by the Numbers appropriate to the Divisions of the Thermometer. The Dryness and Moisture, by the Divisions in the Rim of the Hydroscope. The pressure by Figures, denoting the Height of the Mercurial Cylinder. But for the Faces of the Sky, they are so many, that many of them want proper Names; and therefore it will be convenient to agree upon some determinate ones, by which the most usual may be in brief exprest. As let Clear signify a very clear Sky without any Clouds or Exhalations: Checquer'd a clear Sky, with many great white round Clouds, such as are very usual in Summer. Hazy, a Sky that looks whitish, by Reason of the Thickness of the higher Parts of the Air, by some Exhalations not formed into Clouds. Thick, a Sky more whiten'd by a greater Company of Vapours: these do usually make the Luminaries look bearded or hairy, and are oftentimes the Cause of the Appearance of Rings and Haloes about the Sun as well as the Moon. Overcast, when the Vapours so whiten and thicken the Air that the Sun cannot break through; and of this there are very many Degrees, which may be exprest by a little, much, more, very much overcast, &c. Let Hairy signify a Sky that hath many small, thin, and high Exhalations, which resemble Locks of Hair, or Flakes of Hemp or Flax; whose Varieties may be exprest by strait or curv'd &c. according to the Resemblance they bear. Let Water'd signify a Sky that has many high thin and small Clouds, looking almost like water'd Tabby, call'd in some Places a Mackrel Sky. Let a Sky be call'd Wav'd, when those Clouds appear much bigger and lower, but much after the same manner. Cloudy, when the Sky has many thick dark Clouds. Lowring, when the sky is not very much overcast, but hath also under-neath many thick dark Clouds which threaten Rain. The Signification of gloomy, foggy, misty, fleeting, driving, rainy, snowy, Reaches or Racks variable, &c. are well known, they being very commonly used. There may be also several Faces of the Sky compounded of two or more of these, which may be intelligibly enough exprest by two or more of these Names. It is likewise desirable, that the Particulars of the eighth and ninth Columns may be entred in as little Room and as few Words as are sufficient to signify them intelligibly and plainly.

It were to be wisht that there were diverse in several Parts of the World, but especially in distant Parts of this Kingdom, that would undertake this Work, and that such would agree upon a common Way somewhat after this Manner, that as near as could be, the same Method and Words might be made Use of. The Benefit of which Way is easily enough conceivable.

As for the Method of using and digesting those so collected Observations; that will be more advantageously considered when the *Supellex* is provided; a Workman being then best able to fit and prepare his Tools for his Work, when he sees what Materials he has to work upon.

MARTIN LISTER:
from the *Philosophical Transactions of the Royal Society* (before 1700)

The Reason of the Ascent of the Quicksilver

IT is observed of the Barometer, that the Quick-silver is not affected with the Weather, or very rarely, let that be either cloudy, rainy, windy, or serene, in St Helena, or the Barbadoes: and therefore probably not within the Tropicks, unless in a violent Storm or Hurricane. The first is affirmed by Mr Halley, who kept a Glass near two Months in the Island of St Helena, and the other of Barbadoes stands upon the Credit of our Registers.

In England, in a violent Storm, or when the Quick-silver is at the very lowest, it then visibly breaks and emits small Particles, as I have more than once observed; which Disorder I look upon as a kind of Fretting; and consequently at all Times of its Descent, it is more or less upon the Fret.

In this Disorder of the Quick-silver, I imagine it hath its Parts contracted, and closer put together; which seems probable, for that, for Example, the Quick-silver then emits, and squeezes out fresh Particles of Air into the Tube, which encreasing the Bulk of the Air, and consequently its Elasticity the Quick-silver is necessarily depress'd thereby, that is, by an external Force or Power; and also the Quick-silver must of itself come closer together, in its own internal Parts, that is, descends, for both Reasons.

And that much Air is mixed with it, appears from the Application of a heated Iron to the Tube, as is practised in the purging of it that way; and also for that polish'd Iron will rust, though immers'd in it, as some Philosophers have lately observed.

Now when the Quick-silver rises in the Pipe (which it certainly does both in hot and frosty Weather) it may then be said to be in a natural State, free, open, and expanded like it self, which it seems it ever is within the Tropicks, and with us only in very hot and very frosty Weather. But when it descends, it is then contracted, and as it were convulsed and drawn together, as it mostly is in our Climate of England, and more or less, as we guess, in all Places on this Side the Tropicks. Which Contraction plainly appears from the concave Figure of both Superficies, not only in that of the Quick-silver in the Tube, but also (if well observed) in that which stagnates in the Pot or Dish it self.

The Difficulty seems to lie in the reconciling the same Effect of the Quick-silver's rising in the Tube, from such seemingly different Causes, as great Heat and intense Frost: and those who shall willingly assent to us in one Particular, and grant us Warmth as a probable Cause of its Restitution to its Nature, will yet be at a Stand how to imagine, that great Frost likewise should bring the Quick-silver nearer its own Nature too. I answer, that Salts liquified will coagulate or crystallize, that is, will return to their own proper Natures, both in Cold and in Heat; and therefore, tho' most Men practise the setting them in a cool Cellar for that Purpose, yet some (as Zwelfer) advise, as the best Means to have them speedily and fairly crystallized, to keep them constantly *in Balneo*. Thus also the *Lympha* of the Blood does become a Jelly, if you set it in a cool Place, and the same is by Warmth in like manner inspissated. Again, that it is no new Opinion, that Water is naturally Ice, if no Disquiet from some external Accident hinder. Bornichius the learned Dane has said something for it: And altho' some may think that what he hath said, was a mere Compliment to his own frozen Climate, yet I dare venture to add, in Confirmation of that Doctrine, that Salt is naturally Rock, that is, naturally fossile, not liquid; and yet this is most like Ice of any thing in Nature, not only because of its Transparency, but also for its easy Liquefaction, and the sudden Impressions and Changes which Air makes upon it, so that it is scarce to be preserved in its natural State of Crystallization. Also Salts of all sorts seem naturally to propagate themselves in a hard State, and to vegetate in a dry Form. The like is to be observed in Quick-silver, of its being a hard Rock, and also from its Willingness to embrace upon all Occasions a more fixt State, as in its Amalgamizing with almost all sorts of Metals.

It will not be amiss by way of Corollary, to add a Note or two about healthful and sickly Seasons, more particularly as they may refer to this Phænomenon of great Cold and Frost. If therefore Quick-silver and Liquids are nearest their own Natures, and have less Violence done to them, in very cold and very hot Seasons, the Humours of our Bodies, as Liquids, in all

Probability must be in some measure accordingly affected. And that therefore Cold is healthful, I argue from the vast Number of old Men and Women, to be found upon the Mountains of England, comparatively to what are found elsewhere.

Again, the Blood it self, or the vital Liquor of Animals equivalent to it, is in most Kinds of Animals in Nature sensibly cold; for that the Species of Quadrupeds and Fowls are not to be compared for Number to Fishes and Insects; there being, in all probability, by what I have observed, above a hundred Species of these latter Creatures, whose vital Juice is cold, to one of the former: But because we most converse with those whose vital Juice is hot, we are apt to think the same of all.

Again, I have observed, which I offer as an Argument of the little Injury intense Cold does to the Nature of Animals; I say, I have seen both Hexapode Worms (which I compare to the tender Embryo's of sanguineous Animals, because such are in a middle State) and Flies of divers Sorts, hard frozen in the Winter, and I have taken them up from the Snow, and if I cast them against the Glass, they would endanger the breaking of it, and make it ring like so much hard Ice; yet when I put the Insects under the Glass, and set them before the Fire, they would, after a short Time, nimbly creep about, and be gone, if the Glass which I whelmed upon them, had not secured them.

It hath indeed been noted by a very wise Philosopher in Contradiction to our English Proverb, which says, that A green Christmas makes a fat Churchyard; that the last Plague broke out here at London, after a long and severe Winter 1665. But I reply, That that was accidentally only; for that Disease is never bred amongst us, but comes to us by Trade and Infection. 'Tis properly a Disease of Asia, where it is Epidemical. And therefore, by the Providence of God, we are very secure from any such Calamities as the natural Effect of our Climate. But we are not to judge or prognosticate of the Salubrity or Sickliness of a Year, from foreign Diseases, but the raging of such as are natural to the Men of our Climate.

ROBERT BOYLE:

from *New Experiments Physico-Mechanical touching the Spring of the Air* (1660)

Experiment I

FOR the more easy understanding of the experiments triable by our engine, I thought it not superfluous nor unseasonable in the recital of this first of them, to insinuate that notion, by which it seems likely, that most, if not all of them, will prove explicable. Your Lordship will easily suppose, that the notion I speak of is, that there is a spring, or elastical power in the air we live in. By which ελατὴρ or spring of the air, that which I mean is this: that our air either consists of, or at least abounds with, parts of such a nature, that in case they be bent or compressed by the weight of the incumbent part of the atmosphere, or by any other body, they do endeavour, as much as in them lieth, to free themselves from that pressure, by bearing against the contiguous bodies that keep them bent; and, as soon as those bodies are removed, or reduced to give them way, by presently unbending and stretching out themselves, either quite, or so far forth as the contiguous bodies that resist them will permit, and thereby expanding the whole parcel of air, these elastical bodies compose.

This notion may perhaps be somewhat further explained, by conceiving the air near the earth to be such a heap of little bodies, lying one upon another, as may be resembled to a fleece of wool. For this (to omit other likenesses betwixt them) consists of many slender and flexible hairs; each of which may indeed, like a little spring, be easily bent or rolled up; but will also, like a spring, be still endeavouring to stretch itself out again. For though both these hairs, and the aëreal corpuscles to which we liken them, do easily yield to external pressures; yet, each of them (by virtue of its structure) is endowed with a power or principle of self-dilatation; by virtue whereof, though the hairs may by a man's hand be bent and crouded closer together, and into a narrower room than suits best with the nature of the body; yet, whilst the compression lasts, there is in the fleece they compose an endeavour outwards, whereby it continually thrusts against the hand that opposes its expansion. And upon the removal of the external pressure, by opening the hand more or less, the compressed wool doth, as it were, spontaneously expand or display itself towards the recovery of its former more loose and free condition, till the fleece hath either regained its former dimensions, or at least approached them as near as the compressing hand (perchance not quite opened) will permit. This power of self-dilatation is somewhat more conspicuous in a dry spunge compressed, than in a fleece of wool. But yet we rather chose to employ the latter on this occasion, because it is not, like a spunge, an entire body, but a number of slender and flexible bodies, loosely complicated, as the air itself seems to be.

There is yet another way to explicate the spring of the air; namely, by supposing with that most ingenious gentleman, Monsieur Des Cartes, that the air is nothing but a congeries or heap of small and (for the most part) of flexible particles, of several sizes, and of all kind of figures, which are raised by heat (especially that of the sun) into that fluid and subtle ethereal body that surrounds the earth; and by the restless agitation of that celestial matter, wherein those particles swim, are so whirled round, that each corpuscle endeavours to beat off all others from coming within the little sphere requisite to its motion about its own centre; and in case any, by intruding into that sphere, shall oppose its free rotation, to expel or drive it away: so that, according to this doctrine, it imports very little, whether the particles of the air have the structure requisite to springs, or be of any other form (how irregular soever) since their elastical power is not made to depend upon their shape or structure, but upon the vehement agitation, and (as it were) brandishing motion, which they receive from the fluid æther, that swiftly flows between them, and whirling about each of them (independently from the rest) not only keeps those slender aëreal bodies separated and stretched out (at least, as far as the neighbouring ones will permit) which otherwise, by reason of their flexibleness and weight, would flag or curl; but also makes them hit against, and knock away each other, and consequently require more room than that, which, if they were compressed, they would take up.

By these two differing ways, my Lord, may the springs of the air be explicated. But though the former of them be that, which by reason of its seeming somewhat more easy, I shall for the most part make use of in the following discourse; yet am I not willing to declare peremptorily for either of them against the other. And indeed, though I have in another treatise endeavoured to make it probable, that the returning of elastical bodies (if I may so call them) forcibly bent, to their former position, may be mechanically explicated; yet I must confess, that to determine whether the motion of restitution in bodies proceed from this, that the parts of a body of a peculiar structure are put into motion by the bending of the spring, or from the endeavour of some subtle ambient body, whose passage may be opposed or obstructed, or else its pressure unequally resisted by reason of the new shape or magnitude, which the bending of a spring may give the pores of it: to determine this, I say, seems to me a matter of more difficulty, than at first sight one would easily imagine it. Wherefore I shall decline meddling with a subject, which is much more hard to be explicated than necessary to be so by him, whose business it is not, in this letter, to assign the adequate cause of the spring of the air, but only to manifest, that the air hath a spring, and to relate some of its effects.

I know not whether I need annex, that though either of the above-mentioned hypotheses, and perhaps some others, may afford us an account plausible enough of the air's spring; yet I doubt, whether any of them gives us a sufficient account of its nature. And of this doubt I might here mention some reasons, but that, peradventure, I may (God permitting) have a fitter occasion to say something of it elsewhere. And therefore I should now proceed to the next experiment, but that I think it requisite, first, to suggest to your Lordship what comes into my thoughts, by way of answer to a plausible objection, which I foresee you may make against our proposed doctrine, touching the spring of the air. For it may be alledged, that though the air were granted to consist of springy particles (if I may so speak) yet thereby we could only give an account of the dilatation of the air in wind-guns, and other pneumatical engines, wherein the air hath been compressed, and its springs violently bent by an apparent external force; upon the removal of which, it is no wonder, that the air should, by the motion of restitution, expand itself till it hath recovered its more natural dimensions: whereas, in our above-mentioned first experiment, and in almost all others triable in our engine, it appears not, that any compression of the air preceded its spontaneous dilatation or expansion of itself. To remove this difficulty, I must desire your Lordship to take notice, that of whatever nature the air, very remote from the earth, may be, and whatever the schools may confidently teach to the contrary, yet we have divers experiments to evince, that the atmosphere we live in is not (otherwise than comparatively to more ponderous bodies) light, but heavy. And did not their gravity hinder them, it appears not why the streams of the terraqueous globe, of which our air in great part consists, should not rise much higher, than the refractions of the sun, and other stars, give men ground to think, that the atmosphere, (even in the judgment of those recent astronomers, who seem willing to enlarge its bounds as much as they dare,) doth reach.

But lest you should expect my seconding this reason by experience; and lest you should object, that most of the experiments, that have been proposed to prove the gravity of the air, have been either barely proposed, or perhaps not accurately tried; I am content, before I pass further, to mention here, that I found a dry lamb's bladder, containing near about two thirds of a pint, and compressed by a packthread tied about it, to lose a grain and the eighth part of a grain of its former weight, by the recess of the air upon my having prickt it: and this with a pair of scales, which, when the full bladder and the correspondent weight were in it, would manifestly turn either way with the 32[d] part of the grain. And if it be further objected, that the air in the bladder was violently compressed by the packthread and the sides of the bladder, we might probably (to wave prolix answers) be furnished with a reply, by setting down the differing weight of our receiver, when emptied, and when full of uncompressed air, if we could here procure scales fit for so nice an experiment; since we are informed, that in the German experiment, commended at the beginning of this letter, the ingenious triers of it found, that their glass vessel, of the capacity of 32 measures, was lighter when the air had been drawn out of it, than before, by no less than one ounce and $\frac{3}{10}$, that is, an ounce and very near a third. But of the gravity of the air, we may elsewhere have occasion to make further mention.

SOUND (OR ACOUSTICS)

ROBERT BOYLE:
from *New Experiments Physico-Mechanical touching the Spring of the Air* (1660)

Experiment XXVII

THAT the air is the medium, whereby sounds are conveyed to the ear, hath been for many ages, and is yet the common doctrine of the schools. But this received opinion hath been of late opposed by some philosophers upon the account of an experiment made by the industrious Kircher, and other learned men; who have (as they assure us) observed, that if a bell, with a steel clapper, be so fastened to the inside of a tube, that upon the making the experiment *de vacuo* with that tube, the bell remained suspended in the deserted space at the upper end of the tube: and if also a vigorous load-stone be applied on the outside of the tube to the bell, it will attract the clapper, which, upon the removal of the load-stone falling back, will strike against the opposite side of the bell, and thereby produce a very audible sound; whence divers have concluded, that it is not the air, but some more subtle body, that is the medium of sounds. But because we conceived, that, to invalidate such a consequence from this ingenious experiment, (though the most luciferous that could well be made without some such engine as ours) some things might be speciously enough alledged; we thought fit to make a trial or two, in order to the discovery of what the air doth in conveying of sounds, reserving divers other experiments triable in our engine concerning sounds, till we can obtain more leisure to prosecute them.

Conceiving it then the best way to make our trial with such a noise, as might not be loud enough to make it difficult to discern slighter variations in it, but rather might be, both lasting (that we might take notice by what degrees it decreased) and so small, that it could not grow much weaker without becoming imperceptible; we took a watch, whose case we opened, that the contained air might have free egress into that of the receiver. And this watch was suspended in the cavity of the vessel only by a pack-thread, as the unlikeliest thing to convey a sound to the top of the receiver; and then closing up the vessel with melted plaister, we listened near the sides of it, and plainly enough heard the noise made by the balance. Those also of us, that watched for that circumstance, observed, that the noise seemed to come directly in a streight line from the watch unto the ear. And it was observable to this purpose, that we found a manifest disparity of noise, by holding our ears near the sides of the receiver, and near the cover of it: which difference seemed to proceed from that of the texture of the glass, from the structure of the cover (and the cement) through which the sound was propagated from the watch to the ear. But let us prosecute our experiment. The pump after this being employed, it seemed, that from time to time the sound grew fainter and fainter; so that when the receiver was emptied as much as it used to be for the foregoing experiments, neither we, nor some strangers, that chanced to be then in the room, could, by applying our ears to the very sides, hear any noise from within; though we could easily perceive, that by the moving of the hand, which marked the second minutes, and by that of the balance, that the watch neither stood still, nor remarkably varied from its wonted motion. And to satisfy ourselves farther, that it was indeed the absence of the air about the watch, that hindered us from hearing it, we let in the external air at the stop-cock; and then though we turned the key and stopt the valve, yet we could plainly hear the noise made by the balance, though we held our ears sometimes at two foot distance from the outside of the receiver; and this experiment being reiterated into another place, succeeded after the like manner. Which seems to prove, that whether or no the air be the only, it is at least the principal medium of sounds. And by the way it is very well worth noting, that in a vessel so well closed as our receiver, so weak a pulse as that the balance of a watch, should propagate a motion to the air in a physically streight line, notwithstanding the interposition of so close a body as glass, especially glass of such thickness as that of our receiver; since by this it seems the air imprisoned in the glass must, by the motion of the balance, be made to beat against the concave part of the receiver, strongly enough to make its convex part beat upon the contiguous air, and so propagate the motion to the listner's ears. I know this cannot but seem strange to those, who, with an eminent modern philosopher, will not allow, that a sound, made in the cavity of a room, or other place so closed, that there is no intercourse betwixt the external and internal air, can be heard by those without, unless the sounding body do immediately strike against some part of the inclosing body. But not having now time to handle controversies, we shall only annex, that after the foregoing experiment, we took a bell of about two inches in diameter at the bottom, which was supported in the midst of the cavity of the receiver by a bent stick, which by reason of its spring pressed with its two ends against the opposite parts of the inside of the vessel: in which, when it was closed up, we observed, that the bell seemed to sound more dead than it did when just before it sounded in the open air. And yet, when afterwards we had (as formerly) emptied the receiver, we could not discern any considerable change (for some said they observed a small one) in the loudness of the sound. Whereby it seemed, that though the air be the principal medium of sound, yet either a more subtle matter may be also a medium of it, or else an ambient body, that contains but very few particles of air, in comparison of those it is easily capable of, is sufficient for that purpose.

And this, among other things, invited us to consider, whether in the above-mentioned experiment made with the bell and the load-stone, there might not in the deserted part of the tube remain air enough to produce a sound; since the tubes for the experiment *de vacuo* (not to mention the usual thinness of the glass) being seldom made greater than is requisite, a little air might bear a not inconsiderable proportion to the deserted space: and that also, in the experiment *de vacuo*, as it is wont to be made, there is generally some little air, that gets in from without, or at least store of bubbles, that arise from the body of the quicksilver, or other liquor itself, observations heedfully made have frequently informed us; and it may also appear, by what hath been formerly delivered concerning the Torricellian experiment.

On the occasion of this experiment concerning sounds, we may add in this place, that when we tried the experiment formerly mentioned, of firing gun-powder with a pistol in our evacuated receiver, the noise made by the striking of the flint against the steel was exceeding languid, in comparison of what it would have been in the open air. And on divers other occasions it appeared, that the sounds created within our exhausted glass, if they were not lost before they reached the ear, seemed at least to arrive there very much weakened. We intended to try, whether or no the wire-string of an instrument shut up into our receiver would, when the ambient air was sucked out, at all tremble, if in another instrument held close to it, but without the receiver, a string tuned (as musicians speak, how properly I now examine not) to an unison with it, were briskly touched, and set a vibrating. This, I say, we purposed to try, to see how the motion made in the air without would be propagated through the cavity of our evacuated receiver. But when the instrument, wherewith the trial was to be made, came to be employed, it proved too big to go into the pneumatical vessel: and we have not now the conveniency to have a fitter made.

We thought likewise to convey into the receiver a long

and slender pair of bellows, made after the fashion of those usually employed to blow organs, and furnished with a small musical instead of an ordinary pipe. For we hoped, that by means of a string fastened to the upper part of the bellows, and to the moveable stopple, that makes a part of the cover of our receiver, we should, by frequently turning round that stopple, and the annexed string, after the manner already often recited, be able to lift up and distend the bellows; and by the help of a competent weight fastened to the same upper part of the bellows, we should likewise be able at pleasure to compress them, and, by consequence, try whether, that subtler matter than air (which, according to those that deny a vacuum, must be supposed to fill the exhausted receiver) would be able to produce a sound in the musical pipe; or in a pipe like that of ordinary bellows, to beget a wind capable to turn or set on moving some very light matter, either shaped like the sails of a wind-mill, or of some other convenient form, and exposed to its orifice. This experiment, I say, we thought to make; but have not yet actually made it for want of an artificer to make us such a pair of bellows as it requires.

We had thoughts also of trying, whether or no, as sounds made by the bodies in our receiver become much more languid than ordinary, by reason of the want of air; so they would grow stronger, in case there were an unusual quantity of air crouded and shut up in the same vessel. Which may be done (though not without some difficulty) by the help of the pump, provided the cover and stopple be so firmly fastened (by binding and cement, or otherwise) to the glass and to each other, that there be no danger of the condensed air's blowing of either of them away, or its breaking through the junctures. These thoughts, my Lord, as I was saying, we entertained; but for want of leisure, as, of as good receivers as ours, to substitute in its place, in case we should break it before we learned the skill of condensing the air in it, we durst not put them in practice. Yet on this occasion give me leave to advertise your

Lordship once for all, that though for the reasons newly intimated, we have only in the seventeenth experiment, taken notice, that by the help of our engine the air may be condensed as well as rarefied; yet there are divers other of our experiments, whose phænomena it were worth while to try to vary, by means of the compression of the air.

NARCISSUS MARSH:

from the *Philosophical Transactions of the Royal Society* (1683)

The Doctrine of Sounds

I CANNOT better explain the Usefulness of this Theory of Sounds, than by making a Comparison 'twixt the Faculty of Seeing and Hearing, as to their Improvements. In order to which, I observe, That Vision is threefold, direct, refracted, and reflex'd; answerable whereunto we have Opticks, Dioptricks, and Catoptricks.

In like manner Hearing may be divided into direct, refracted and reflex'd; whereto answer three Parts of our Doctrine of Acousticks, which are yet nameless, unless we call them Acousticks, Diacousticks, and Catacousticks (or in another Sense, but to as good Purpose) Phonicks, Diaphonicks, and Cataphonicks.

Direct Vision has been improved two ways.

1. *Ex parte Objecti*, by the Arts of producing, conserving, and imitating, and duly applying, Light and Colours.

2. *Ex parte Organi vel Medii*, by making use of Tubes without Glasses, or, a Man's closed Hand to look thro'. So likewise direct Hearing, partly has, and partly may further receive great and notable Improvements, both *ex parte Objecti*, and *ex parte Organi vel Medii*.

1. As to the Object of Hearing, which is Sound, Improvement has been and may be made, both as to the begetting, and as to the conveying and propagating (which is a kind of Conserving) of Sounds.

1. As to the begetting of Sounds. The Art of imitating any Sound whether by speaking (that is pronouncing) any kind of Language (which really is an Art; and the Art of Speaking perhaps one of the greatest) or by Whistling, or by Singing (which are allowed Arts) or by Hollowing, or Luring, (which the Huntsman and Falconer would have to be an Art also) or by imitating with the Mouth (or otherwise) the Voice of any Animal; as of Quails, Cats, and the like; or by representing any Sound begotten by the Collision of solid Bodies, or after any other manner; these are all Improvements of direct Hearing, and may be improved.

Moreover, the Skill to make all sorts of musical Instruments, both antient and modern, whether Wind Instruments or string'd, or of any other sort, whereof there are very many (as Drums, Bells, the *Systrum* of the Egyptians, or the like) that Beget (and not only Propagate) Sounds: The Skill of making these, I say, is an Art, that has much improv'd direct Hearing; and an harmonious Sound exceeds a single and rude one, that is an immusical Tune: which Art is yet capable of farther Improvement. And I hope, That by the Rules which may happily be laid down, concerning the Nature, Propagation, and Proportion or adapting of Sounds, a way may be found out, both to improve Musical Instruments already in use, and to invent new ones, that shall be more sweet and luscious than any yet known. Besides, that by the same means Instruments may be made, that shall imitate any Sound in Nature, that is not Articulate; be it of Bird, Beast, or what thing else soever.

2. The Conveying and Propagating (which is a kind of Conserving) of Sounds, is much helped by duly Placing the Sonorous Body, and also by the Medium.

For if the Medium be Thin and Quiescent, and the sounding Body placed conveniently, the Sound will be easily and regularly propagated and mightily conserved.

1. The Medium must be Thin and Quiescent: Hence in a still Evening, or the Dead of the Night (when the Wind

ceases) a Sound is better sent out, and to a greater Distance, than otherwise.

2. The sonorous Body must be placed conveniently, *viz.* Near a smooth Wall, either Plain or Arched (Cycloidically or Elliptically, rather than otherwise; tho a Circular or any Arch will do; but not so well.) Hence in a Church, the nearer the Preacher stands to the Wall (and certainly it's much the best way to place Pulpits near the Wall) the better is he heard, especially by those who stand near the Wall; also, tho at a greater Distance from the Pulpit, those at the remotest End of the Church, by laying their Ears somewhat close to the Wall, may hear him easier than those in the middle.

Hence also do arise Whispering Places. For the Voice being applied to one end of an Arch, easily rowls to the other. And indeed were the Motion and Propagation of Sounds but rightly understood, 'twould be no hard matter to contrive Whispering Places of infinite variety and use. And perhaps there could be no better or more pleasant hearing a Concert of Music, than at such a Place as this; where the Sounds rowling long together before they come to the Ear, must needs consolidate and imbody in one; which becomes a true Composition of Sounds, and is the very Life and Soul of Concert.

2. If the sonorous Body be placed near Water, the Sound will easily be convey'd, yet mollify'd; as Experience teacheth us from a Ring of Bells near a River, and a great Gun shot off at Sea; which differ much in the strength, and yet Softness and Continuance, or Propagation of their Sounds, from the same at Land; where the Sound is more harsh and more perishing, or much sooner decays.

3. In a Plain a Voice may he heard at a far greater Distance, than in uneven Ground. The Reason of all which last nam'd Phænomena is the same; because the sonorous Air meeting with little or no Resistance upon a Plane (much less upon an Arch'd) smooth Superficies, easily rowls along it, without being let or hindred in its Motion, and consequently without

having its Parts disfigured, and put into another kind of Revolution, than what they had at the first begetting of the Sound, which is the true Cause of its Preservation or Progression; and fails much when the Air passes over an uneven Surface, according to the degrees of its Inequality; and somewhat also, when it passes over the Plain Superficies of a Body that is hard and resisting.

Wherefore the smooth Top of the Water (by reason of its yielding to the Arched Air, and gently rising again with a kind of Resurge, like to Elasticity tho it be not so; by which Resurge it quickens and hastens the Motion of the Air rowling over it, and by its yielding preserves it in its Arched Cycloidical or Elliptical Figure) the smooth Top of the Water, I say, for these Reasons, and by these Means, conveys a Sound more entire, and to a greater Distance, than the Plane Surface of a piece of Ground, a Wall, or any other solid Body whatever, can do.

2. The Organ, which is the Ear, is helpt much by placing it near a Wall, (especially at one end of an Arch, the Sound being begotten at the other) or near the Surface of Water, or of the Earth; along which the Sounds are most easily and naturally conveyed; as was before declared. And 'tis Incredible, how far a Sound made upon Earth (by the Trampling of a Troop of Horses, for Example) may be heard in a still Night, if a Man lays his Ear close to the Ground in a large Plain.

Otacousticks here come in for helping the Ear; which may be so contrived (by a right understanding the Progression of Sounds, which is the principal Thing to be known for the due regulating all such kinds of Instruments) as that the Sound might enter the Ear without any Refraction.

2. Refracted Vision (which is always made *ex parte Medii*) arises from the different Density, Figure, and Magnitude of the Medium; which is somewhat altered by the divers Incidence of the Visive Rays, and so it is in Refracted Hearing, all these Causes concur to its Production; and some others to be hearafter considered.

Now as any Object (a Man for Example) seen thro' a Thickened Air, by Refraction appears greater than really he is: So likewise a Sound heard thro' the same Thickened part of the Atmosphere, will be considerably vary'd from what it would seem to be, if heard thro' a Thinner Medium. And this I call a Refracted Sound.

Improvements of Refracted Vision have been made, by Grinding or Blowing Glasses into a certain Figure, and placing them at due Distances; whereby the Object may be (as 'twas) enabled to send forth its Rays more Vigourously, and the Visive Faculty impowered the better to receive them. Thus,

1. A fine Glass Bubble, filled with clear Water, and placed before a burning Candle or Lamp, does help it to dart forth its Rays to a prodigious Length and Brightness.

2. The Visive Faculty is much helped,

1. By Spectacles and other Glasses, which are made to help the Purblind and Weak Eyes, to see at any competent Distance.

2. By Perspective Glasses and Telescopes, which help the Eye to see Objects at a very great Distance, which otherwise would not be discernible.

3. By Microscopes or Magnifying Glasses, which help the Eye to see Near Objects, that by reason of their Smallness were Invisible before.

4. By Polyscopes or Multiplying Glasses, whereby one thing is represented to the Eye as many, whether in the same or different Shapes.

After the same manner, Instruments may be contrived for assisting both the Sonorous Body to send forth its Sound more strongly, and the Acoustrick Faculty to receive and discern it more easily and distinctly. And thus;

1. An Instrument may be invented, that applied to the Mouth (or any Sonorous Body) shall send forth the Voice distinctly as to a prodigious Distance and Loudness. For if the Stentoro-phonicon (which is but a Rude and Inartificial Instrument) does such great Feats; what might be done with One

composed according to the Rules of Art, whose Make should comply with the Laws of Sonorous Motion, which that does not?

2. There are some Instruments, and more such may be Invented to help the Ear; As,

1. Otacousticks (and better may be made) to help Weak Ears to hear at a reasonable Distance also. Which would be as great a help to the Infirmity of Old Age, as the other Invention of Spectacles is, and perhaps greater; for as much as the Hearing what's spoken is of more daily Use and Concern to such Men, than to be able to read Books, or to view Pictures.

2. A sort of Otacousticks may be so contrived, as that they shall receive in Sounds made at a very great Distance, which otherwise would have been Inaudible: And these Otacousticks, in some Respects, would be of greater use than Perspectives.

1. In Time of War, for discovering the Enemy at a good Distance, when he marches or lies incamp'd behind a Mountain or Wood, or any such Place of Shelter, which hinder the Sight from reaching very far.

2. At Sea, when in dark Hazy Weather the Air is too thick, or in Stormy Tempestuous Weather, the Waves rise too high, for the Perspective to be made use of.

3. In Dark Nights, when Perspectives become almost insignificant, and yet at such times, generally, Soldiers take their March, when they would surprize their Enemies.

4. Microphones, or Micracousticks, that is, Magnifying Ear Instruments, which may be contrived after that manner, that they shall render the most Minute Sound in Nature distinctly Audible by Magnifying it to an unconceivable Loudness: By the help whereof we may hear the different Cries and Tones of the smallest Animals.

5. A Polyphone, or Polycoustick, so ordered that One Sound may be heard, either of the Same, or a Different Note: Insomuch that who uses this Instrument, he shall at the Sound of a Single Viol seem to hear a whole Concert, and all True

Harmony. By which means this Instrument has much the Advantage of the Polyscope.

I have called it Refracted Hearing, because made thro' a Medium, *viz*. Thick Air, or an Instrument, thro' which the Sound passing is broken or Refracted.

3. Reflected Vision (which is always made *ex parte Objecti*) hath been improv'd by the Invention of Looking Glasses and Polish'd Metals, whether Plane, Concave or Convex, of several Figures, and placed at Determinate Distances.

In like manner reflex'd Audition (which is only made *ex parte Corporis Oppositi*) may be improved by contriving several sorts of Artificial Ecchoes. For (speaking in general) any Sound falling directly or obliquely upon any dense Body of a smooth (whether Plane or Arch'd) Superficies, is beat back again and reflected, or does Eccho, more or less.

I say (1) Falling directly or obliquely; because, if the Sound be sent out and propagated Parallel to the Surface of the dense Body, there will be no Reflection of Sound, no Eccho.

I say (2) Upon a Body of a smooth Superficies; because if the Surface of the *Corpus Obstans* be uneven, the Air by Reverberation will be put out of its regular Motion, and the Sound thereby broken and extinguish'd: So that, tho in this case also the Air be beaten back again, yet Sound is not reflected, nor is there any Eccho.

I say (3) It does Eccho more or less, to shew, that when all things are, as is before describ'd, there is still an ecchoing, tho it be not always heard, either because the direct Sound is too weak to be beaten quite back again to him that made it; or that it does return home to him, but so weak, that without the help of a good Otacoustick it cannot be discerned; or that he stands in a wrong Place to receive the reflected Sound, which passes over his Head, under his Feet, or to one side of him; which therefore may be heard by a Man standing in that place, where the reflected Sound will come provided no interpos'd Body does intercept it; but not by him that first made it.

These Ecchoes (like Reflected Vision) may be several ways produced, as;

1. A Plane *Corpus Obstans* reflects the Sound back in its due Tone and Loudness; if allowance be made for the proportionable Decrease of the Sound according to its Distance.

2. A Convex *Corpus Obstans* repels the Sound (insensibly) smaller: but somewhat quicker (tho weaker) than otherwise it would be.

3. A Concave *Corpus Obstans* ecchoes back the Sound (insensibly) bigger, slower, (tho stronger) and also inverted; but never according to the order of Words. Nor do I think it possible for the Art of Man to contrive a single Eccho, that shall invert the Sound and repeat backwards; because then the Words last spoken, that is, which do last occur to the *Corpus Obstans*, must first be repell'd; which cannot be. For where in the mean time should the first Words hang and be conceal'd or lie dormant? Or how, after such a Pause be reviv'd and animated again into Motion? Yet in complicated or compound Ecchoes, where many receive from one another, I know not whether something that way may not be done.

From this determinate Concavity or Archedness of these reflecting Bodies, it comes to pass, that some of them from a certain Distance or Positure, will eccho back but one determinate Note, and from no other Place will they reverberate any; because of the undue Position of the sounding Body. Such an one (as I remember) is the Vault in Merton College in Oxford.

4. The Echoing Body, being removed farther off, reflects more of the Sound, than when nearer. And this is the Reason, why some Echoes repeat but one Syllable, some one Word, and some many.

5. Echoing Bodies may be so contriv'd and plac'd, as that reflecting the Sound from one to the other, either directly and mutually, or obliquely and by Succession, out of one Sound shall many Echoes be begotten; which in the first Case will be

all together and somewhat involv'd or swallow'd up of each other; and thereby confused (as a Face in Looking-Glasses obverted;) in the other they will be distinct, separate and succeeding one another, as most multiple Echoes do.

Moreover, a multiple Echo may be made, by so placing the Echoing Bodies, at unequal Distances, that they reflect all one way, and not one on the other; by which means a manifold successive Sound will be heard (not without Astonishment;) one Clap of the Hands like many; one Ha like a Laughter; one single Word like many of the same Tone and Accent; and so one Viol like many of the same kind, imitating each other.

Furthermore, Echoing Bodies may be so ordered, that from any one Sound given, they shall produce many Echoes different both as to their Tone and Intention. By this means a musical Room may be so contrived, that not only one Instrument, played on in it, shall seem many of the same Sort and Size; but even a Concert of (somewhat) different ones; only by placing certain Echoing Bodies so, as that any Note (played) shall be returned by them in 3rds, 5ths and 8ths, which is possible to be done otherwise than was mentioned before in refracted audition.

I have been thus large, that I might give you a little Prospect into the Excellency and Usefulness of Acousticks, and that thereby I might excite others to bend their Thoughts, towards the making of Experiments for the compleating this (yet very Imperfect tho Noble) Science; a Specimen whereof I will give in these three Problems.

Prob. I. To make the least Sound (by the help of Instruments) as loud as the greatest; a Whisper to become as loud as the Shot of a Cannon.

By the help of this Problem the most minute Sounds in Nature may be clearly and distinctly heard.

Prob. II. To propagate any (the least) Sound to the greatest Distance.

By the help hereof any Sound may be conveyed to any and

therefore heard at any Distance, (I must add, within a certain, tho very large Sphere).

Moreover by this means a Weather-Cock may be so contrived, as that with an ordinary blast of Wind it shall cry (or whistle) loud enough to be heard many Leagues. Which haply may be found of some Use, not only for Pilots in mighty tempestuous Weather, when Light Houses are rendred almost useless: But also for the measuring the Strength of Winds, if allowance be made for their different Moisture. For I conceive, That the more dry any Wind is, the louder it will whistle, *cæteris paribus*; I say, *cæteris paribus*, because, besides the strength and dryness of Winds or Breath, there are a great many other things (hereafter to be considered) that concur to the increase of magnifying of Sounds, begotten by them in an Instrument exposed to their Violence, or blown into.

Prob. III. That a Sound may be convey'd from one extreme to the other (or from one distant Place to another) so as not to be heard in the middle.

By the Help of this Problem a Man may talk to his Friend at a very considerable Distance, so that those in the middle Space shall hear nothing of what passed betwixt them.

I shall here add a Semiplane of an Acoustick or Phonical Sphere, as an Attempt to explicate the great Principle in this Science, which is the Progression of Sounds.

You are to conceive this (rude) Semiplane as Parallel to the Horizon; for, if it be Perpendicular thereunto, I suppose the upper Extremity will be no longer Circular, but Hyperbolical, and the lower part of it suited to a greater Circle of the Earth. So that the whole Phonical Sphere (if I may so call it) will be a solid Hyperbola, standing upon a Concave Spherical Base. I speak this concerning Sounds made (as usually they are) nigh the Earth, and whose sonorous Medium has a free passage every way. For if they are generated high in the Air, or directed one way, the Case will be different; which is partly designed in the Inequality of the Draught.

LIGHT (OR OPTICS)

ROBERT HOOKE:
from *Micrographia* (1665)

Of the Colours Observable in Muscovy Glass and other Thin Bodies

MUSCOVY-GLASS, or *Lapis specularis*, is a body that seems to have as many Curiosities in its fabrick as any common mineral I have met with; for first, it is transparent to a great thickness. Next, it is compounded of an infinite number of thin flakes joined or generated one upon another so close and smooth as with many hundreds of them to make one smooth and thin plate of a transparent flexible substance, which with care and diligence may be slit into pieces so exceedingly thin as to be hardly perceivable by the eye, and yet even those, which I have thought the thinnest, I have with a good Microscope found to be made up of many other plates, yet thinner, and it is probable, that, were our microscope much better, we might much further discover its divisibility. Nor are these flakes only regular as to the smoothness of their surfaces, but thirdly, in many plates they may be perceived to be terminated naturally with edges of the figure of a rhomboeid. This figure is much more conspicuous in our English talk, much whereof is found in the Lead Mines and is commonly called spar and Kanck, which is of the same kind of substance with the *Selenitis*, but is seldom found in so large flakes as that is, nor is it altogether so tough, but is much more clear and transparent, and much more curiously shaped, and yet may be cleft and flaked like the other *Selenitis*. But fourthly, this stone has a property, which in respect of the microscope, is more notable and that is that it exhibits several appearances of colours both to the naked eye, but much more conspicuous to the microscope, for the exhibiting of which I took a piece of muscovy-glass and splitting or cleaving it into thin plates, I found that up and down in several parts of them I could plainly perceive several white specks or flaws and others diversly coloured with all the colours of the rainbow and with the microscope I could perceive that these colours were ranged in rings that incompassed the white speck or flaw and were round or irregular, according to the shape of the spot which they terminated, and the position of colours in respect to one another, was the very same as in the rainbow. The consecution of those colours from the middle of the spot outward being blue, purple, scarlet, yellow, green; blue, purple, scarlet and so onwards, sometimes half a score times repeated, that is, there appeared six, seven, eight, nine or ten several coloured rings or lines, each encircling the other, in the same manner as I have often seen a very vivid rainbow to have four or five several rings of colours, that is, accounting all the gradations between red and blue for one. But the order of the colours in these rings was quite contrary to the primary or innermost rainbow and the same with those of the secondary or outermost rainbow; these coloured lines or irises as I may so call them were some of them much brighter than others and some of them also very much broader, they being some of them ten, twenty, nay, I believe near a hundred times broader than others and those usually were broadish which were nearest the centre or middle of the flaw. And oftimes I found that these colours reached to the very middle of the flaw, and then there appeared in the middle a very large spot, for the most part, all of one colour, which was very vivid and all the other colours encompassing it, gradually ascending and growing narrower towards the edges, keeping the same order as in the secondary rainbow, that is, if the middle were blue, the next encompassing it would be a purple, the third a red, the fourth a yellow &c. as above. If the middle were a red, the next without it would be a yellow, the third a green, the fourth a blue and so onward. And this order it always kept whatsoever were the middle colour.

HEAT

EDMUND HALLEY:

from the *Philosophical Transactions of the Royal Society* (*c.* 1700)

The Expansion of several Fluids, in order to ascertain the Divisions of the Thermometer

SINCE the same Degree of Heat does not proportionally expand all Fluids; some swelling with a gentle Warmth, and others not till they be considerably hot; some boiling with a moderate Heat, and others not at all; some capable of great Expansion, others increasing very little; it may well be concluded, that no one of them does increase and diminish in the same Proportion with the Heat, and consequently, that the Thermometers graduated by equal Parts of the Expansion of any Fluid, are not sufficient Standards of Heat or Cold.

This will be more evident from the Experiments which I made in the Months of Feb. and Mar. about 4 Years since (the Weather being reasonably cold and not freezing) with Water, Mercury, and Spirit of Wine; wherein the following Particulars were very remarkable.

I took a large Bolt-head, holding about $3\frac{1}{2}$ *lib.* of Water, with a narrow Neck to make the Augment thereof more sensible; and having filled it with Water, and some few Inches up the Neck; I noted exactly to what Mark the Water came: Then I immersed it into a Skillet of warm Water, and let it stand so long, till I concluded the warm Water had communicated its Temper to the Water included in the Bolt-head; and I found, that though the Water were warm, much beyond the Degree of the Summer's Heat, and notwithstanding it was Winter, yet that gentle Heat had scarce any Effect in dilating the Water; so that it scarce appeared to have ascended in the Neck of the Bolt-head. Then I took the Skillet, and set it over the Fire; when it was observable, that the Water, as it grew hot, did slowly ascend in the Neck, especially at first; but after it began to boil in the Skillet, the Expansion thereof became more visible, and it ascended apace, till such Time as it stopped again; the utmost Effort of boiling Water being able to raise it no higher. Then having made a Mark at the utmost Height whereto it had arisen, I took it out, and had the Satisfaction to observe, that though it was not raised so high without a very strong boiling, yet it subsided very slowly, as retaining some Time the Space it had acquired from the Heat, ever after the Heat was pass'd and the Glass was so cool as to be touched without burning the Fingers. However, the next Morning I found it reduced to the first Mark, where it stood when at first put in, having lost nothing sensible by Evaporation, during the Experiment; which I attribute to the Length of the Neck, wherein the Vapours were condensed into Drops before they reached the Top. Then I examin'd how much Water would raise that in the Neck, to the Mark whereto it had been encreased by boiling, and found it was a 26th Part of the Bulk of the first Water; which, upon repeated Experiments, I found to be true; but it was obvious, that Water increasing so very little, with all the Degrees of Heat the Air receives from the Sun, was a very improper Fluid to make a Thermometer withal; and besides, any freezing Liquor is useless for this Purpose in these Northern Climates.

I took a smaller Bolt-head, with a proportional Cane or Neck, and filled it after the same Manner with Mercury; and having boiled it, as above, I observed that 125 Ounces of Mercury had increased the Space of 810 Grains, or a 74th Part of its Bulk when cold. But it was very remarkable, that whereas a gentle Heat had scarce any Effect on Water, here, on the contrary, the Mercury did sensibly ascend at first, and had almost attainted its greatest Expansion before the Water boiled in the Skillet. And after it boiled, tho' I let it stand very long over the Fire, I could not discern that the most vehement boiling had any Effect on it, above what appeared when it first began to boil. The Mercury being taken out, as it cooled, subsided, and in a few Hours returned to the Mark whereat it stood before it was put into the Water. This Fluid being so sensible of a gentle Warmth, and withal, not subject to evaporate without a good Degree of Fire, might most properly be applied to the Construction of Thermometers, were its Expansion more considerable.

However, small as it is, it is sufficient to disturb the precise Nicety of the Mercurial Barometers shewing the Counterpoise of the Pressure of the Atmosphere by a Cylinder of Mercury: For if Mercury be more expanded, and consequently lighter in warm Weather than in cold; it will necessarily follow, that the same Weight of Atmosphere will require a taller Cylinder in Summer, and a shorter in Winter to counterpoise it. And if the Extremity of Weather do but occasion a 15th Part of Difference, as 'tis probable it doth, the Effect thereof, on a Barometer, will be a Tenth of an Inch above the Mean, or a Fifth in all.

I fill'd the smaller Bolt-head with Spirit of Wine; and having set it in the Skillet of Water over the Fire, I found that it ascended gradually, as the Heat increased, but slower at first, and faster after it was well warm. At length being arrived at a certain Degree of Heat, it would fall a boiling with great Violence, emitting Bubbles, which coming into the Neck of the Bolt-head, would lift all the incumbent Spirits till they had made their Way through. And these succeeding one another very fast, would often raise the Spirit to the Top of the Neck, and spill it; so that I found I could go no further with this Liquor, than to that Degree of Heat which occasioned this boiling, and which wanted very much of that of boiling Water, being almost tolerable to the Touch. It was however very remarkable, how exactly this Degree of Heat was determined by the Expansion of the Spirit; for in the Instant it reached a certain Mark on the Neck, it began to emit its Bubbles: And having been taken out a little to cool and subside, it would certainly and constantly fall a bubbling again, when upon a second Immersion, it was arrived at the foresaid Mark. During this Experiment, it appear'd both by the Dew on the Neck, and by the Scent in the Room, that tho' the Neck were about 30 Inches long, yet the Spirit did evaporate very fast for the Smallness of the Surface of the Liquor: And I have often noted the like Evaporations condensed in Dew within the Head of the ordinary seal'd Thermometers, in very hot Weather.

This Degree of Heat which made the Spirit of Wine begin to boil, being determined so nicely as I have said, made me conclude, that this might very properly be taken for the Limit of the Scale of Heat in a Thermometer; and the Effect thereof in the Expansion of any other Fluid being accurately noted, might be easily transferr'd to any sort of Thermometer whatsoever. Only it must be observed, that the Spirit of Wine used to this purpose, be highly rectify'd or dephlegmed; for otherwise the differing Goodness of the Spirit will occasion it to boil sooner or later, and thereby pervert the designed Exactness. And by the way, give me Leave to hint, that the sooner or later boiling of Spirits or spirituous Liquors may possibly be as good a Test of their Strength and Perfection, as their specifick Gravity, or any other yet used.

The Spirit of Wine I made use of was possibly none of the best; but I observed, that at the Point of boiling it had increased a 12th Part in bulk; which great Dilatation makes it a Liquor sufficiently adapted to our Purpose, were it not for the Evaporation thereof, and for the Difference in Goodness of the Spirit, and for that, in Length of Time it becomes as it were Effete, and loses gradually a Part of its expansive Power.

This expansive Power is in no Fluid comparably so conspicuous as in that rare elastick Fluid the Air; for by several

Experiments that I have made, I find that the Heat of Summer does expand the ordinary Air about a 30th Part; and that late honourable Patron of experimental Philosophy, Mr Boyle, alledges his own Trials, proving that the Force of the strongest Cold in England does not contract the Air above 1/20 Part. So that the same Air, which in extreme Cold occupies 12 Parts of Space, in very hot Summer Weather, will require 13 such Spaces; which is as great an Expansion as that of Spirit of Wine when it begins to boil: For which Reason, and for its being so very sensible of Warmth and Cold, and continuing to exert the same elastick Power, after never so long being included, in my Opinion, it is much the most proper Fluid for the Purpose of Thermometers.

Now the Thermometers hitherto in Use, are of two Sorts: the one shewing the different Temper of Heat and Cold, by the Expansion of the Spirit of Wine, the other by the Air: But I cannot learn that either of them of either Sort, were ever made or adjusted, so as it might be concluded, what the Degrees or Divisions of the said Instruments did mean; neither were they ever otherwise graduated, but by Standards kept by each particular Workman, without any Agreement or Reference to one another: So that whatsoever Observations any curious Person may make by his Thermometer, to signify the Degree of Heat in the Air, or other Things (which is of constant Use in Philosophical Matters) cannot be understood, unless by those who have by them Thermometers of the same Make and Adjustment. Much less has the Way been shewn how to make this Instrument without a Standard, or to make two of them agree artificially, without comparing them together.

I shall only add, that whereas the usual Thermometers with Spirit of Wine, do some of them begin their Degrees from a Point, which is that whereat the Spirit stands when it is so cold as to freeze Oil of Anniseeds; and others from the Point of beginning to freeze Water: I conceive these Points are not so justly determinable, but with a considerable Latitude: And that the just Beginning of the Scales of Heat and Cold should not be from such a Point as freezes any Thing, but rather from Temperature, such as is in Places deep under Ground, where the Heat of the Summer, or Cold in Winter, have (by the certain Experiment of the curious M. Mariotte, in the Grottoes under the Observatory at Paris) been found to have no Manner of Effect.

CHEMISTRY AND GENERAL PROPERTIES OF MATTER

Robert Boyle:
from *The Sceptical Chymist* (1661)

A Dialogue on the Nature of Combustion

You recall to my mind (says Carneades) a certain experiment I once devised, innocently to deceive some persons, and let them and others see, how little is to be built upon the affirmation of those, that are either unskilful or unwary, when they tell us they have seen alchymists make the mercury of this or that metal: and to make this the more evident, I made my experiment much more slight, short and simple, than the chymists usual processes to extract metalline mercuries; which operations being commonly more elaborate and intricate, and requiring a much more longer time, give the alchymists a greater opportunity to cozen, and confrequently are more obnoxious to the spectators suspicion. And that, wherein I endeavoured to make my experiment look the more like a true analysis, was, that I not only pretended, as well as others, to extract a mercury from the metal I wrought upon, but likewise to separate a large proportion of manifest and inflammable sulphur. I take then of the filings of copper about a dram or two; of common sublimate, powdered, the like weight; and sal armoniac near about as much as of sublimate: these three being well mingled together, I put into a small phial with a long neck, or, which I find better, into a glass-urinal, which (having first stopped it with cotton) to avoid the noxious fumes, I approach by degrees to a competent fire of well kindled coals, or (which looks better, but more endangers the glass) to the flame of a candle; and after a while the bottom of the glass being held just upon the kindled coals, or in the flame, you may in about a quarter of an hour, or, perchance, in half that time, perceive in the bottom of the glass some running mercury; and if you then take away the glass, and break it, you shall find a parcel of quick-silver, perhaps altogether, and perhaps part of it in the pores of the solid mass. You shall find too, that the remaining lump being held to the flame of the candle, will readily burn with a greenish flame, and after a little while (perchance presently) will in the air acquire a greenish blue, which being the colour that is ascribed to copper, when its body is unlocked, it is easy to persuade men, that this is the true sulphur of *Venus*, especially since not only the salts may be supposed partly to be flown away, and partly to be sublimed to the upper part of the glass, (whose inside will commonly appear whitened by them) but the metal seems to be quite destroyed, the copper no longer appearing in a metalline form, but almost in that of a resinous lump: whereas, indeed, the case is only this, that the saline parts of the sublimate, together with the sal armoniac, being excited and actuated by the vehement heat, fall upon the copper, (which is a metal they can more easily corrode than silver) whereby the small parts of the mercury being freed from the salts, that kept them asunder, and being by the heat tumbled up and down after many occursions, they convene into a conspicuous mass of liquor; and as for the salts, some of the more volatile of them subliming to the upper part of the glass, the others corrode the copper, and uniting themselves with it, do strangely alter and disguise its metallick form, and compose with it a new kind of concrete, inflammable like sulphur; concerning which, I shall not now say any thing, since I can refer you to the diligent observations, which I remember Mr Boyle has made concerning this odd kind of verdigrease. But, continues Carneades smiling, you know I was not cut out for a mountebank, and therefore I will hasten to resume the person of a sceptick, and take up my discourse, where you diverted me from prosecuting it.

In the next place then I consider, that, as there are some bodies, which yield not so many as the three principles; so there are many others, that in their resolution exhibit more principles than three; and that therefore the ternary number is not that of the universal and adequate principles of bodies. If you allow of the discourse I lately made you, touching the primary associations of the small particles of matter, you will scarce think it improbable, that of such elementary corpuscles there may be more sorts than either three, or four, or five. And if you will grant, what will scarce be denied, that corpuscles of a compounded nature may, in all the wonted examples of chymists pass for elementary, I see not why you should think it impossible, that *aqua fortis*, or *aqua regis*, will make a separation of colliquated silver and gold, though the fire cannot: so there may be some agent found out so subtile and so powerful, at least in respect of those particular compounded corpuscles, as to be able to resolve them into those more simple ones, whereof they consist, and consequently increase the number of the distinct substances, whereunto the mixed body has been hitherto resoluble. And if that be true, which I recited to you a while ago out of *Helmont*, concerning the operations of the alkahest, which divides bodies into other distinct substances, both as to number and nature, than the fire does; it will not a little countenance my conjecture. But confining ourselves to such ways of analyzing mixed bodies, as are already not unknown to chymists, it may without absurdity be questioned, whether besides those grosser elements of bodies, which they call salt, sulphur and mercury, there may not be ingredients of a more subtile nature, which being extremely little, and not being in themselves visible, may escape unheeded at the junctures of the distillatory vessels, though never so carefully luted. For let me observe to you one thing, which, though not taken notice of by chymists, may be a notion of good use in divers cases to a naturalist, that we may well suspect, that there may be several sorts of bodies, which are not immediate objects of any one of our senses; since we see, that not only those little corpuscles, that issue out of the load-stone, and perform the wonders, for which it is justly admired; but the effluviums of amber, jet, and other electrical concretes, though by their effects upon the particular bodies disposed to receive their action, they seem to fall under the cognizance of our sight, yet do they not as electrical immediately affect any of our senses, as do the bodies, whether minute or greater, that we see, feel, taste, &c. But (continues Carneades) because you may expect I should, as the chymists do, consider only the sensible ingredients of mixt bodies, let us now see what experience will, even as to these, suggest to us.

ROBERT BOYLE:

from *New Experiments Physico-Mechanical touching the Spring of the Air* (1660)

Experiment XXXV

THIS occasion, I have had to take notice of siphons, puts me in mind of an odd kind of siphon, that I caused to be made a pretty while ago; and which hath been since, by an ingenious man of your acquaintance, communicated to divers others. The occasion was this: an eminent Mathematician told me one day, that some inquisitive Frenchmen (whose names I know not) had observed, that in case one end of a slender and perforated pipe of glass be dipped in water, the liquor will ascend to some height in the pipe, though held perpendicular to the plain of the water. And, to satisfy me, that he misrelated not the experiment, he soon after brought two or three small pipes of glass, which gave me the opportunity of trying it; though I had the less reason to distrust it, because I remember I had often, in the long and slender pipes of some weather-glasses, which

I had caused to be made after a somewhat peculiar fashion, taken notice of the like ascension of the liquor, though (presuming it might be casual) I had made but little reflection upon it. But after this trial, beginning to suppose, that though the water in these pipes, that were brought me, rise not above a quarter of an inch (if near so high) yet, if the pipes were made slender enough, the water might rise to a very much greater height; I caused several of them to be, by a dexterous hand, drawn out at the flame of a lamp, in one of which, that was almost incredibly slender, we found, that the water ascended (as it were of itself) five inches by measure, to the no small wonder of some famous Mathematicians, who were spectators of some of these experiments. And this height the water reached to, though the pipe were held in as erected a posture as we could; for if it were inclined, the water would fill a greater part of it, though not rise higher in it. And we also found, that when the inside of the pipe was wetted beforehand, the water would rise much better than otherways. But we caused not all our slender pipes to be made streight, but some of them crooked, like siphons: and having immersed the shorter leg of one of these into a glass, that held some fair water, we found, as we expected, that the water arising to the top of the siphon, though that were high enough, did of itself run down the longer leg, and continue running like an ordinary siphon. The cause of this ascension of the water appeared to all that were present so difficult, that I must not stay to enumerate the various conjectures that were made at it, much less to examine them; especially having nothing but bare conjectures to substitute in the room of those I do not approve. We tried indeed, by conveying a very slender pipe and a small vessel of water into our engine, whether or no the exsuction of the ambient air would assist us to find the cause of the ascension we have been speaking of: but though we employed red wine instead of water, yet we could scarce certainly perceive thorough so much glass, as was interposed betwixt our eyes and the liquor, what happened in a pipe so slender, that the redness of the wine was scarce sensible in it. But, as far as we could discern, there happened no great alteration to the liquor; which seemed the less strange, because the spring of that air, that might depress the water in the pipe, was equally debilitated with that, which remained to press upon the surface of the water in the little glass. Wherefore, in favour of his ingenious conjecture, who ascribed the phænomenon under consideration to the greater pressure made upon the water by the air without the pipe, than by that within it, (where so much of the water, consisting perhaps of corpuscles more pliant to the internal surfaces of the air, was contiguous to the glass) it was shown, that in case the little glass-vessel, that held the water, of which a part ascended into the slender pipe, were so closed, that a man might with his mouth suck the air out of it, the water would immediately subside in the small pipe. And this would indeed infer, that it ascended before only by the pressure of the incumbent air; but that it may (how justly I know not) be objected, that peradventure this would not happen, in case the upper end of the pipe were in a vacuum; and that it is very probable the water may subside, not because the pressure of the internal air is taken off by the exsuction, but by reason of the spring of the external air, which impels the water, it finds in its way to the cavity deserted by the other air, and would as well impel the same water upwards, as make it subside, if it were not for the accidental posture of the glasses. However, having not now leisure to examine any farther this matter, I shall only mind your Lordship, that if you will prosecute this speculation, it will be pertinent to find out likewise, why the surface of water (as is manifest in pipes) useth to be concave, being depressed in the middle, and higher on every side: and why in quick-silver, on the contrary, not only the surface is wont to be very convex, or swelling, in the middle; but if you dip the end of a slender pipe in it, the surface of the liquor (as it is called) will be lower within the pipe, than without. Which phænomena, whether, and how far, they may be deduced from the figure of the mercurial corpuscles, and the shape of the springy particles of the air, I willingly leave to be considered.

Robert Boyle:
from *The History of Fluidity* (1661)

On Fluids

For the third and chief condition of a fluid body is, that the particles it consists of be agitated variously and apart, whether by their own innate and inherent motion, or by some thinner substance, that tumbles them about in its passage through them. For this seems to be the main difference betwixt solid ice and fluid water, that in the one the parts (whether by any newly-acquired texture, or for want of sufficient heat to keep them in motion) being at rest against one another, resist those endeavours of our fingers to displace them, to which in the other, the parts being already in motion, easily give way. For whereas in the ice, every part actually at rest must by the law of nature continue so, till it be put out of it by an external force capable to surmount its resistance to a change of its present state; in water each corpuscle being actually (though but slowly) moved, we need not begin or produce a new motion in it, but only byass or direct that, which it has already, which many familiar instances manifest to be a much easier task. From this agitation of the small parts of liquors it comes to pass, that these little bodies, to continue their motion, do almost incessantly change places, and glide sometimes over, sometimes under, and sometimes by the sides of one another. Hence also may be rendered a reason of the softness of fluid bodies, that is, their yielding to the touch: for the particles that compose them being small, incoherent, and variously moved, it can be no difficult matter (as we lately intimated) to thrust them out of those places, which being already in motion they were disposed to quit, especially there being vacant rooms at hand, ready to admit them as soon as they are displaced. And hence it likewise happens, that these little bodies must be very easily moveable any way upon the motion of the mass or liquor, which they compose; and that being very small, and moving so many ways, they cannot but (according to Aristotle's definition of things fluid) be very unfit to bound themselves, but very easy to be bounded by any other firm body; for that hinders them from spreading any further: and yet to continue their various and diffusive motion as much as they can, (especially their gravity, at least here about the earth, equally depressing and thereby levelling as to sense their uppermost superficies) they must necessarily move to and fro, till their progress be stopped by the internal surface of the vessel, which by terminating their progress (or motion toward the same part) does consequently necessitate the liquor those little bodies compose, to accommodate itself exactly (for aught the eye is able to discern to the contrary) to its own figure.

EDMUND HALLEY:

from the *Philosophical Transactions of the Royal Society* (1686)

The Effects of Gravity in the Descent of Heavy Bodies and the Motion of Projects

DES CARTES his Notion, I must needs confess to be to me incomprehensible, while he will have the Particles of his Cælestial Matter, by being reflected on the Surface of the Earth, and so ascending therefrom, to drive down into their Places those Terrestrial Bodies they find above them: This is, as near as I can gather, the Scope of the 20, 21, 22 and 23 Sections of the last Book of his *Principia Philosophiæ*; yet neither he nor any of his Followers can shew, how a Body suspended *in Libero Æthere*, shall be carried downwards by a continual Impulse tending upwards, and acting upon all its Parts equally: And besides, the Obscurity wherewith he expresses himself, particularly, Sect. 23. does sufficiently argue, according to his own Rules, the Confused Idea he had of the thing he wrote.

Others, and among them, Dr Vossius, assert the Cause of the Descent of Heavy Bodies, to be the Diurnal Rotation of the Earth upon its Axis, without considering that according to the Doctrine of Motion fortified with Demonstration, all Bodies moved *in Circulo*, would recede from the Center of their Motion; whereby the contrary to Gravity would follow, and all loose Bodies would be cast into the Air in a Tangent to the Parallel of Latitude, without the Intervention of some other Principle to keep them fast, such as is that of Gravity. Besides the Effect of this Principle is thro'out the whole Surface of the Globe found nearly equal, and certain Experiment seems to argue it rather less near the Equinoctal, than towards the Poles; which could not be by any means, if the Diurnal Rotation of the Earth upon its Axis were the Cause of Gravity; for where the Motion was swiftest, the Effect would be most considerable.

Others assign the Pressure of the Atmosphere to be the Cause of this Tendency towards the Center of the Earth; but unhappily they have mistaken the Cause for the Effect, it being from undoubted Principles plain, that the Atmosphere has no other Pressure, but what it derives from its Gravity; and that the Weight of the upper Parts of the Air, pressing on the lower Parts thereof, do so far bend the Springs of that Elastick Body, as to give it a Force equal to the Weight that compressed it, having of itself no Force at all: And supposing it had, it will be very hard to explain the *Modus*, how that Pressure should occasion the Descent of a Body circumscribed by it, and pressed equally above and below, without some other Force to Draw or Thrust it downwards. But to demonstrate the contrary of this Opinion, an Experiment was long since shewn before the Royal Society; whereby it appeared, that the Atmosphere was so far from being the Cause of Gravity, that the Effects thereof were much more vigorous where the Pressure of the

Atmosphere was taken off; for a long Glass-Receiver, having a light Down-Feather included, being Evacuated of Air, the Feather, which in the Air would hardly sink, did *in Vacuo* descend with nearly the same Velocity as if it had been a Stone.

Some think to illustrate this Descent of Heavy Bodies, by comparing it with the Virtue of the Loadstone; but setting aside the Difference there is in the manner of their Attractions, the Loadstone drawing only in and about its Poles, and the Earth near equally in all Parts of its Surface, this Comparison avails no more than to explain *Ignotum per æque Ignotum.*

Others assign a certain Sympathetical Attraction between the Earth and its Parts, whereby they have, as it were, a Desire to be united, to be the Cause we enquire after: But this is so far from explaining the Modus, that it is little more, than to tell us in other terms, that Heavy Bodies descend, because they descend.

But tho the Efficient Cause of Gravity be so obscure, yet the Final Cause thereof is clear enough; for it is by this single Principle that the Earth and all the Celestial Bodies are kept from Dissolution: the least of their Particles not being suffered to recede far from their Surfaces, without being immediately brought down again by Virtue of this Natural Tendency, which, for their Preservation, the Infinite Wisdom of their Creator has ordained to be towards each of their Centers; nor can the Globes of the Sun and Planets otherwise be destroyed, but by taking from them this Power of keeping their Parts united.

The Affections or Properties of Gravity, and its manner of acting upon Bodies Falling, have been in a great measure discovered, and most of them made out by Mathematical Demonstration in this our Century, by the accurate Diligence of Galilæus, Torricellius, Hugenius, and others; and now lately, by our worthy Countryman Mr Is. Newton. Which Properties I shall here enumerate.

The First Property is, That by this Principle of Gravitation, all Bodies do descend towards a Point, which either is, or else is very near to the Center of Magnitude of the Earth and Sea, about which the Sea forms itself exactly into a Spherical Surface, and the Prominences of the Land, considering the Bulk of the whole, differ but insensibly therefrom.

MAGNETISM

EDMUND HALLEY:

from the *Philosophical Transactions of the Royal Society* (*c.* 1692)

A Theory of the Magnetical Variation

BEFORE I proceed to the Theory of the Variation of the Magnetical Compass, it is necessary to lay down the Grounds upon which I raise my Conclusions; and at once to give a Synopsis of those Variations which I have reason to look upon as sure, being mostly the Observations of persons of good Skill and Integrity.[1]

By this Table it appears,

First, That in all Europe the Variation at this Time is West, and more in the Eastern Parts thereof than the Western; as likewise that it seems throughout to be upon the Increase that way.

2. That on the coast of America, about Virginia, New England. and Newfoundland, the Variation is likewise westerly; and that it increases all the way as you go northerly along the Coast, so as to be above 20 deg. at Newfoundland, nearly 30 deg. in Hudson's Streights, and not less than 57 deg. in Baffin's Bay; also that as you sail Eastwards from this Coast, the Variation diminishes. From these two it is a Legitimate Corollary, That somewhere between Europe and the North Part of America, there ought to be an easterly Variation, or at least no westerly; and so I conjecture it's about the eastermost of the Tercera Islands.

3. That on the Coast of Brasil there is East Variation, which increases very notably as you go to the Southward, so as to be 12 deg. at Cape Frio, and over against the River of Plata 20½

[1] A table of variations is inserted here in the original.

deg. and from thence, sailing South westerly to the Streights of Magellan, it decreases to 17 deg. and at the West Entrance it is but 14 deg.

4. That to the Eastward of Brasil properly so called, this easterly Variation decreases, so as to be very little at St Helena and Ascension; and to be quite gone, and the Compass to point true about 18 deg. of Longitude, West from the Cape of Good Hope.

5. That to the Eastward of the aforesaid Places, a Westward Variation begins, which reigns in the whole Indian Sea, and arises to no less than 18 deg. under the Æquator itself, about the Meridian of the Northern Part of Madagascar, and near the same Meridian; but in 39 Deg. South Latitude, it is found full 27½ deg. from thence easterly, and West Variation decreases so as to be but little more than 8 deg. at Cape Comorin, and then 3 deg. upon the Coast of Java, and to be quite extinct about the Molucca Islands, as also a little to the Westwards of Van Diemen's Land, found out by the Dutch in 1642.

6. That to the Eastward of the Molucca's and Van Diemen's Land in South Lat. there arises another easterly Variation, which seems not so great as the former, nor of so large Extent; for that at the Island Roterdam it is sensibly less than upon the East-Coast of New Guinea: and at the rate it decreases, it may well be supposed, that about 20 deg. farther East, or 25 deg. East Long. from London, in the Latitude of 20 deg. South, a westerly Variation begins.

7. That the Variation observed by the Hon. Sir John Narborough, at Baldivia, and at the West Entrance of the Streights of Magellan, do plainly shew, that the East Variation noted in our 3d Remark is decreasing apace, and that it cannot reasonably extend many Degrees into the South Sea from the Coast of Peru and Chili, leaving room for a small westerly Variation in that Tract of the unknown World, that lies in the Mid-way between Chili and New Zealand, and between Hound's Island and Peru.

8. That in sailing North-west from St Helena, by Ascension, as far as the Æquator, the Variation continues very small East, and as it were constantly the same: So that in this part of the World, the Course wherein there is no Variation, is evidently no Meridian, but rather Northwest.

9. That the Entrance of Hudson's Streights and the Mouth of the River of Plata, being nearly under the same Meridian, at the one Place the Needle varies 29½ deg. to the West, at the other 20½ deg. to the East. This plainly demonstrates the Impossibility of reconciling these Variations by the Theory of Bond; which is by two Magnetical Poles and an Axis, inclin'd to the Axis of the Earth; from whence it would follow that under the same Meridian, the Variation should be in all Places the same way.

These things being premised, may serve as a sure Foundation for this Theory. That the whole Globe of the Earth is one great Magnet, having 4 Magnetical Poles, or Points of Attraction, near each Pole of the Æquator, and that in those Parts of the World which lie near adjacent to any of those Magnetical Poles, the Needle is governed thereby; the nearest Pole being always predominant over the more remote. The Parts of the Earth wherein these Magnetical Poles lie, cannot as yet be exactly determined for want of sufficient Data to proceed geometrically: But as near as Conjecture can reach, I reckon that the Pole which is at present nearest to us, lies in or near the Meridian of the Land's End of England, and not above 7 deg. from the Pole Arctick. By this Pole the Variations in all Europe and Tartary, and the North Sea, are principally governed, yet with regard to the other Northern Pole, whose Situation is in a Meridian passing about the middle of California, and about 15 deg. from the North Pole of the World. To this the Needle has chiefly respect in all the North-America, and in the two Oceans on either Side thereof, from the Azores westwards to Japan, and farther. The two Southern Poles are rather farther distant from the South Pole of the World: The one about 16 deg. therefrom, is in a Meridian some 20 deg. to the Westward of Magellan's Streights, or 95 deg. West from London; this commands the Needle in all the South-America, in the Pacifick Sea, and the greatest part of the Ethiopic Ocean. The 4th and last Pole seems to have the greatest Power and largest Dominions of all, as it is the most remote from the Pole of the World, being little less than 20 deg. distant therefrom, in the Meridian which passes through Hollandia nova, and the Island of Celebes, about 120 deg. East from London. This Pole is predominant in the South part of Africa, in Arabia, and the Red Sea, in Persia, India, and its Islands, and all over the Indian Sea from the Cape of Good Hope Eastwards to the middle of the great South-Sea that divides Asia from America. This seems to be the present Disposition of the Magnetical Virtue throughout the whole Globe of the Earth.

By this Hypothesis it is plain that (our European North Pole being in the Meridian of the Land's End of England) all Places more Easterly than that will have it on the West-side of the Meridian; and consequently the Needle respecting it with its Northern Point, will have a Westerly Variation, which will still be greater as you go to the Eastwards, till you come to some Meridian of Russia, where it will be greatest, and from thence decrease again. Thus at Brest the Variation is but $1\frac{1}{3}$ deg. at London $4\frac{1}{2}$ deg. but at Dantzick 7 deg. West. To the Westward of the Meridian of the Land's End, the Needle ought to have an Easterly Variation, were it not that (by approaching the American Northern Pole, which lies on the West-side of the Meridian, and seems to be of greater Force than this other) the Needle is drawn thereby Westward, so as to counterbalance the Direction given by the European Pole, and to make a small West Variation in the Meridian of the Land's End itself. Yet I suppose that about the Meridian of the Isle Tercera, our nearest Pole may so far prevail as to give the Needle a little turn to the East, though but for a very small Space, the Counterbalance of those two Poles permitting no considerable Variation in all the Eastern Parts of the Atlantick Ocean, nor upon the West Coasts of England and Ireland, France, Spain and Barbary. But to the Westwards of the Azores, the power of the American Pole overcoming that of the European, the Needle has chiefly respect thereto; and turns still more and more towards it as you approach it. Whence it comes to pass, that on the Coast of Virginia, New-England, Newfoundland, and in Hudson's Streights, the Variation is Westwards; that it decreases as you go from thence towards Europe; and that it is less in Virginia and New-England than in Newfoundland, and Hudson's Streights. This Westerly Variation again decreases, as you pass over the North America; and about the Meridian of the Middle of California, the Needle again points due North; and from thence Westwards to Yedzo, and Japan, I make no doubt but the Variation is Easterly; and half Sea over not less than 15 deg. This East Variation extends over Japan, Yedzo, Tartary, and part of China, till it meet with the Westerly, which is governed by the European North-Pole, and which I said was greatest somewhere in Russia.

Towards the Southern-Pole the Effect is much the same, only that here the South point of the Needle is attracted. Hence it will follow, that the Variation on the Coast of Brazil, at the River of Plata, and so on to the Streights of Magellan, should be Easterly, as in the 3d Remark. And this Easterly Variation doth extend Eastward over the greatest part of the Ethiopick Sea, till it be counterpoised by the Virtue of the Southern-Pole; as it is about mid-way between the Cape of Good Hope, and the Isles of Tristan d' Alcunha. From thence Eastwards the Asian South-Pole (as I must take the liberty to call it) becoming prevalent, and the South point of the Needle being attracted thereby, there arises a West Variation very great in Quantity and Extent, because of the great Distance of this Magnetical Pole from the Pole of the World. Hence it is, that in all the Indian Sea as far as Hollandia nova, and farther, there is constantly West Variation: And that under the Equator itself, it arises to no less than 11 deg. where it is most. About the Meridian of the Island Celebes, being likewise that of this Pole, this Westerly Variation ceases, and an Easterly begins; which reaches, according to my Hypothesis, to the Middle of the South-Sea between Zelandia nova and Chili, leaving room for a small West Variation governed by the American South-Pole; which I shewed to be in the Pacifick Sea, in the 6th and 7th Remarks.

In the Torrid Zone, and particularly under the Equinoctial, respect must be had to all 4 Poles, and their Positions well considered; otherwise it will not be easy to determine what the Variation shall be, the nearest Pole being always the strongest; yet not so, as not to be counterbalanced sometimes by the united Forces of two more remote. A notable Instance hereof is in our 8th Remark, where I took notice, that in sailing from St Helena, by the Isle of Ascension, to the Equator on a N.W. Course, the Variation is very little Easterly, and in that whole Tract unalterable: For which I give this Reason, that the South American Pole (which is considerably the nearest in the aforesaid Places) requiring a great Easterly Variation, is counterpoised by the contrary attraction of the North American and the Asian South-Poles; each whereof singly is, in these Parts, weaker than the American South-Pole: And upon the N.W. Course, the Distance from this latter is very little varied; and as you recede from the Asian South-Pole, the Balance is still preserved by the Access towards the North American Pole. I mention not in this Case the European North-Pole, its Meridian being little removed from those of these Places, and of itself requiring the same Variations we here find.

What I have here said does plainly shew the sufficiency of this Hypothesis, solving the Variations that are at this time observed.

But there are two Difficulties not easy to surmount. The one is, That no Magnet I have ever seen or heard of, hath more than two opposite Poles: Whereas the Earth hath visibly four, and perhaps more. Secondly, It is plain by the change of the Variation, not only at London, where this Discovery was first made, but also almost all over the Earth, that these Poles are not, at least all of them, fixed in the Earth, but shift from Place to Place, whereas it is not known that the Poles of the Load-stone ever shifted their Place in the Stone, nor, considering the compact Hardness of that Substance, can it easily be supposed. These Difficulties for a long time made me despond, when in accidental Discourse, and least expecting it, I stumbled on the following Hypothesis.

It is sufficiently known and allowed, that the Needle's Variation changes; and that this Change is gradual and universal, will appear by the following Examples.

At London, An.1580. The Variation was observed by Mr Burrows, to be 11 deg. 15 min. East. In An.1622, the same was found by Mr Gunter, to be but 6 deg. East. In the Year 1634, Mr Gellibrand found it 4 deg. 5 min. East. In 1657, Mr Bond observed that there was no Variation at London. An.1672, myself observed it 2 deg. 30 min. to the West, and this present Year 1692, I again found it 6 deg. West. So that in 112 Years the Direction of the Needle has changed no less than 17 Degrees.

At Paris, Orontius Finæus about the Year 1550, did account it about 8 or 9 deg. East Variation. An.1640, it was found 3 deg. East. An.1666, there was no Variation there, and An. 1681, I found it to be 2 deg. 3 min. to the West.

At Cape d'Agulhas, the most Southerly Promontory of Africa, about the Year 1600, the Needle pointed due North and South without Variation, whence the Portugueze gave its Name. An.1622, there was 2 deg. West Variation. An.1675, it was 8 deg. West: And this Year 1692, it was curiously observed not less than 11 deg. West.

At St Helena, about the Year 1600, the Needle declined 8 deg. to the East. An.1623, It was but 6 deg. East. An.1677, when I was there, I observed it accurately on Shore to be 40 min. East; and now this Year it was found 1 deg. to the Westward of the North.

At Cape Comorin in India, in the Year 1620, there was 14 deg. 20 min. West Variation; in the Year 1680, there was 8 deg. 48 min. but in the year 1688, it was no more than 7 deg. 30 min. so that here the Needle has returned to the East, about 7 deg. in 70 Years.

From these, and many other Observations, it is evident that the Direction of the Needle is in no Place fixed and constant, though in some it changes faster than in others. And where for a long it has continued as it were unaltered, it is there to be understood, that the Needle has its greatest Deflection, and is become Stationary, in order to return, like the Sun in the Tropick. This at present is in the Indian Sea, about the Island Mauritius, where is the highest West Variation and in a Tract tending from thence into the N.N.W. towards the Red Sea and Egypt. And in all Places to the Westward of this Tract, all over Africa and the Seas adjoining, the West Variation will be found to have increased; to the Eastwards thereof, as in the Example of Cape Comorin, to have decreased, *viz.* all over the East-Indies, and the Islands near it.

After the like manner, in that Space of East Variation, which, beginning near St Helena, is found all over the South America, and which at present is highest about the Mouth of Rio de la Plata, it has been observed, that in the Eastern Parts thereof the Variation of the Needle gradually decreases. And by Analogy we may infer, though we have not Experience enough to ascertain it, that both the East and West Variation in the Pacifick Sea, do gradually increase and decrease after the same Rule.

These Phænomena being well understood, and duly considered, do sufficiently evince, that the whole Magnetical System is by one, or perhaps more Motions translated: That this moving thing is very great, as extending its Effects from Pole to Pole; and that the Motion thereof is not *per saltum*, but by a gradual and regular Motion.

TECHNOLOGY

SIR WILLIAM PETTY:
from Thomas Sprat's *History of the Royal Society* (1667)

An Apparatus to the History of the Common Practices of Dying

IT were not incongruous to begin the History with a Retrospect into the very nature of Light it self (as to inquire whether the same be a Motion or else a Body;) nor to premise some Theorems about the Sun, Flame, Glow-worms, the Eyes of some Animals, shining Woods, Scales of some Fishes, the dashing of the Sea, strokes upon the Eyes, the Bolonian Slate (called by some the Magnet of Light) and of other light and lucid Bodies.

It were also not improper to consider the very essentials of Colour and Transparencies (as that the most transparent Bodies, if shaped into many angles, present the eye with very many colours;) That bodies having but one single superficies, have none at all, but are suscipient of every colour laid before them; That great depths of Air make a Blue, and great depths of Water a Greenish colour; That great depths or thicknesses of coloured Liquors do all look blackish (red Wine in a large conical Glass being of all reddish colours between Black at the top and White at the bottom).

That most Vegetables, at one time or other, are greenish; and that as many things passing the Sun are blackned, so many others much whitened by the same: Other things are whitened by acid Fumes, as red Roses and raw Silks by the smoak of Brimstone.

Many Metals, as Steel and Silver, become of various colours, and tarnish by the Air, and by several Degrees of heat.

We might consider the wonderful variety of colours appearing in Flowers, Feathers; and drawn from Metals, their Calces and Vitrifications; and of the Colours rising out of transparent Liquors artificially mixed.

But these things, relating to the abstracted nature of Colours, being too hard for me, I wholly decline; rather passing to name (and but to name) some of the several sorts of Colorations now commonly used in Humane affairs, and as vulgar Trades in these Nations; which are these; *viz*.

1. There is a whitening of Wax, and several sorts of Linnen and Cotton Cloths, by the Sun, Air, and by reciprocal effusions of Water.

2. Colouring of Wood and Leather by Lime, Salt, and Liquors, as in Staves, Canes, and Marble Leather.

3. Colouring of Paper, *viz*. Marbled Paper, by distempering the colours with Ox-gall, and applying them upon a stiff gummed Liquor.

4. Colouring, or rather discolouring the Colours of Silks, Tiffanies, &c. by Brimstone.

5. Colouring of several Iron and Copper-work, into Black, with Oyl.

6. Colouring of Leather into Gold-colour, or rather Silver leaves into Gold by Varnishes, and in other cases by Urine and Sulphur.

7. Dying of Marble and Alabaster with heat and coloured Oyls.

8. Colouring Silver into Brass with Brimstone or Urine.

9. Colouring the Barrels and Locks of Guns into Blue and Purple with the temper of Small-coal heat.

10. Colouring of Glass (made of Sands, Flints, &c.) as also of Chrystals and Earthen Ware, with the rusts and solutions of Metals.

11. The colouring of live Hair, as in Poland, Horse and Man's Hair; as also the colouring of Furrs.

12. Enameling and Anealing.

13. Applying Colours, as in the Printing of Books and Pictures, and as in making of playing Cards; being each of them performed in a several way.

14. Gilding and Tinning with Mercury, Block-Tin, Sal-Armoniack.

15. Colouring Metals, as Copper with Calamy into Brass, and with Zink or Spelter into Gold, or into Silver with Arsenick: And of Iron into Copper with Hungarian Vitriol.

16. Making Painter's Colours by preparing of Earth, Chalk, and Slates; as in Umber, Oker, Cullen earth, &c. as also out of Calces of Lead, as Ceruse and Minium; by Sublimates of Mercury and Brimstone, as in Vermilion; by tinging of white Earths variously, as in Verdeter, and some of the Lakes; by concrete Juices or *Fæculæ*, as in Gambrugium, Indico, Pinks, Sap-green, and Lakes: As also by Rusts, as in Verdegreese, &c.

17. The applying of these Colours by the adhesion of Ox-gall, as in the Marble Paper aforesaid; or by Gum-water, as in Limning; or by clammy drying Oyls, (such as are the Oyls of Linseed, Nuts, Spike, Turpentine, &c.)

18. Watering of Tabbies.

19. The last I shall name is the colouring of Wool, Linnen, Cotton, Silk, Hair, Feathers, Horn, Leather, and the Threads and Webbs of them with Woods, Roots, Herbs, Seeds, Leaves, Salts, Limes, Lixiviums, Waters, Heats, Fermentations, Macerations, and other great variety of Handling: An account of all which is that History of Dying we intend. All that we have hitherto said being but a kind of remote and scarce pertinent Introduction thereunto.

I begin this History by enumerating all the several Materials and Ingredients which I understand to be or to have been used in any of the last aforementioned Colorations, which I shall represent in various Methods, *viz*. out of the Mineral Family. They use Iron and Steel, or what is made or comes from them, in all true Blacks (called Spanish Blacks) though not in Flanders Blacks; *viz*. they use Copperas, Steel-filings, and Slippe, which is the stuff found in the Troughs of Grind-stones, whereon Edge-tools have been ground. They also use Pewter for Bow-dye, Scarlet, *viz*. they dissolve Bars of Pewter in the *Aqua-fortis* they use; and make also their Dying-kettles or Furnace of this Metal.

Litharge is used by some, though acknowledged by few, for what necessary reason I cannot learn, other than to add weight unto Dyed Silk; Litharge being a Calx of Lead, one of the heaviest and most colouring Metals.

I apprehend Antimony much used to the same purpose, though we know there be a very tingent Sulphur in their Mineral, which affordeth variety of Colour by the precipitations and other operations upon it.

Arsenick is used in Crimson upon pretence of giving Lustre, although those who pretend not to be wanting in giving Lustre to their Silks, do utterly disown the use of Arsenick.

Verdegrease is used by Linnen Dyers in their Yellow and Greenish Colours, although of itself it strike not deeper Colour than of Pale Straws.

Of Mineral Salts used in Dying; the chief is Allum; the very true use thereof seems to me obscure enough, notwithstanding all the Narrations I could get from Dyers about it: For I doubt,

1. Whether it be used to make common Water a fit Menstruum, wherewith to extract the Tingent particles of several hard Materials; for I find Allum to be used with such Materials as spend easy enough, as Brasil, Logwood, &c. And withal, that the Stuffs to be dyed are first boyled in Allum-Liquors, and the Allum afterwards (as they say) cleared from the said Stuff again, before any Colour at all be apply'd.

2. Whether it be used to scour the Sordes, which may interpose between the *Coloranda*, and the Dying Stuff; and so hinder the due adhesion of the one unto the other: The boyling of several things first in Allum seeming to tend this way. But I find this work to be done in Cloth, and Rugs, by a due scouring

of the same in the Fulling-mills with Earth, and in Silk with Soaps, by which they boyl out the Gums and other Sordes, hindring or vitiating the intended Colours.

3. Whether Allum doth intenerate the Hairs of Wool, and Hair-Stuffs, as Grograins, &c. Whereby they may the better receive and imbibe their Colours? Unto which opinion I was led by the Dyers; saying, that after their stuffs were well boyled in Allum, that they then cleared them of the Allum again: But we find the most open bodied Cottons and Silks, to have Allum used upon them; as well as the harder Hairs. Nor is Allum used in many Colours, *viz.* in no Woad or Indico Blues; and yet the Stuffs dyed Blue, are without any previous inteneration quickly tinged; and that with a sight and short immersion thereof into the Blue-fat.

4. Whether it contribute to the Colour it self, as Copperas doth to Galls, in order to make a Black; or as Juice of Lemons doth to Cocheneel in the Incarnadives; or as Aquafortis impregnated with Pewter, doth in the Bow-Scarlet, changing it from a red Rose-Crimson to Flame-Colour. This use is certainly not to be denied to Allum in some cases; but we see in other cases, that the same Colours may be dyed without Allum, as well as with it, though neither so bright and lively, nor so lasting.

5. Wherefore, Fifthly, I conclude (as the most probable opinion) that the use of Allum is to be a Vinculum between the Cloth and the Colour, as clammy Oyls and Gum-waters are in Painting and Limning; Allum being such a thing, whose particles and Aculei dissolved with hot Liquors will stick to the Stuffs, and pitch themselves into their Pores; and such also, as on which the particles of the dying Drugs will also catch hold, as we see the particles of Copperas and other chrystallizing materials do of Boughs and Twigs in the Vessel, where such Chrystallization is made. A second use I imagine of Allum in Dying, to be the extracting or drying up of some such particles, as could not consist with the Colour to be super-

induced; for we see Allum is used in the dressing of Alutas or white Leather, the which it dryeth, as the Salt of Hen-dung doth in Ox-hides, and as common Salt doth in preservation of Flesh-meats; for we know, a Sheep-skin newly slayed could not be colour'd as Brasils are, unless it were first dressed into Leather with Allum, &c. which is necessary to the Colour, even although the Allum be, as it is, cleared out of the Leather again, before the said Colouration, with Bran, Yolks of Eggs, &c. Wherefore as Allum, as it were by accident, makes a wet raw Skin to take a bright Colour, by extracting some impedimental particles out of it; so doth it also out of other materials, though perhaps less discernably.

Another use I suppose of Allum, which is to brighten a Colour: For as we see the finest and most glassy materials to make the most orient Colours, as Feathers, Flowers, etc. so certainly if by boyling Cloth in Allum, it becomes incrustated with particles, as it were of Glass, the tinging of them yields more brightness, than the tinging of a Scabrous matter (such as unallumed Cloth is) can do. Analogous hereunto I take the use of Bran, and Bran-liquors in Dying to be; for Bran yielding a most fine Flour (as we see in the making of white Starch;) I conceive that this Flour entring into the pores of the Stuff, levigates their Superficies, and so makes the Colour laid on it, the most orient Colours, as Feathers, Flowers, &c. so certainly be gilded, are first smoothned over with white Colours, before the Gold be laid on.

And indeed all other Woods are filled, not only as to their greater holes and Asperities, with Putty; but also their smaller *Scabrities* are cured by priming Colours before the ultimate Colour intended be laid thereon.

The next Mineral Salt is Salt-peter, not used by ancient Dyers, and but by few of the modern. And that not till the wonderful use of *Aqua-fortis* (whereof Salt-peter is an ingredient) was observed in the Bow-scarlet: Nor is it used now, but to brighten Colours by back-boyling them; for which use

Argol is more commonly used. Lime is much used in the working of Blue-fats, being of Lime-stone calcined, and called Calke, of which more hereafter.

Thomas Henshaw:
from Thomas Sprat's *History of the Royal Society* (1667)

The Manner of Making Salt-peter

In the first place you must be provided of eight or ten Tubs, so large, that they may be able to contain about ten Barrows full of Earth each of them. These Tubs must be all open at the top; but in the bottom of every one of them, you must make a hole near to that side you intend to place outermost, which hole you must fit very well with a Tap and Spigot on the outside downward. On the inside of the Tub, near the Tap-hole, you must carefully place a large wad of straw, and upon that a short piece of board, which is all to keep the earth from stopping up the Tap-hole. When you have placed your Tubs on their stands, at such a distance one from the other, that you may come with ease between them, then fill them up with such *Peter-earth* as you have chosen for your work, leaving only void about a span's breadth between the earth and the edge of the Tub; then lay on the top of the Earth in each Tub, as near as you can to the middle, a rundle of Wicker, like the bottom of a Basket, and about a foot in diameter, and by it stick into the Earth a good strong Cudgel, which must be thrust pretty near the bottom; the Wicker is to keep the water, when it is poured on, from hollowing and disordering the Earth, and the Cudgel is to be stirred about, to give the Water ingress to the Earth upon occasion: Then pour on your Earth common cold Water, till it stand a hand's breadth over the Earth: When it has stood eight or ten hours, loosen the Spigots, and let the Water rather dribble, than run into half Tubs, which must be set under the Taps: This *Lixivium* the Workmen call their Raw-liquor; and note, that if it come not clear at the first drawing, you must pour it on again, and after some little time draw it off, till it come clear, and of the colour of Urine.

If you are curious to know how rich your Liquor is before boyling, you may take a Glass-phial, containing a Quart; fill it with the common Water you use, then weigh it exactly; next fill the same Glass with your Liquor, and find the difference of weight, which compared with the quantity of all your Liquors, will give you a very near guess, how much Salt-peter you are like to make by that boyling.

Then pour on again, on the same Earth, more common Water, that it may bring away what is remaining in the Earth of the former Liquor. This second Liquor is of no other use, but to be poured on new Earth, instead of common Water, because it contains some quantity of Salt-peter in it.

When this is done, turn out the useless insipid Earth out of the Tubs, which you must fill with new Earth, and continue this Operation, till you have in the same manner lixiviated all the Earth: Then fill your Copper with your Liquor, which Copper, for one of the Profession, must be about two hundred weight, and set strongly in a Furnace of brick-work; besides, on one side of your Furnace, you are to place a Tub full of your Liquor, which at a Tap below may dribble as fast into the Copper, as the force of the Fire doth waste your Liquor, which Invention is only to save charges in Fewel. When you have boyled it up to that height, that a little of it flirted off the finger on a live Charcoal, will flash like Gun-powder (which for the most part falls out to be about two Days and a Nights boiling) at what time, upon tryal, a hundred weight of the Liquor contains about five and thirty pound weight of *Peter*. But the Workmen seldom make use of any further indication, than by finding the Liquor hang like Oyl on the sides of the brasen Scummer, when 'tis dipped into it, which is a sign it is fit to be passed through the Ashes, which is done in this manner.

You must prepare two Tubs, fitted after the manner of the

first, where you put your Earth, saving that at the bottom of these Tubs, you must lay Reeds or Straw a foot high; over them place loose boards, pretty near one another; over them, a little more Straw (which is to keep the Ashes from the top, to give the Liquor room to drein the better from them:) Then fill up your Tubs with any sort of Wood-ashes to half a foot of the top; then pour on the foresaid Liquor, as it comes scalding hot out of the Copper, on the Ashes contained in the first Tub; then after a while draw it off at the top; and so continue putting on and drawing off, first at one Tub of Ashes, then at the other, till your Liquor grow clear, and lose the thick turbid colour it had when it went on.

When all the Liquor hath in this manner past through the Ashes of both Tubs, that by this means all its greasy oyl is left behind in the Ashes, you must keep it for the second boyling in a Vessel by itself; in the mean time pour upon your Ashes a sufficient quantity of common water, very hot, once or twice, to bring away what is remaining of the Liquor in the Ashes.

When you begin the second boyling, put first into the Copper the Water that went last through your Ashes, and as that wasteth, let your strong Liquor drop into the Copper, out of the Tub above described, standing on the side of the Furnace, till the Liquor in the Copper be ready to shoot or chrystalise.

Note, That toward the end of your boyling, there will arise great store of Scum and Froth, which must be carefully taken off with a great brass Scummer, made like a Ladle, full of little holes, and usually about that time it lets fall some common Salt to the bottom, which you must take up with the said Scummer, and lay it aside for another use.

To know when the Liquor is ready to shoot into *Peter*, you need but drop a little of it on a knife, or any other cold thing that hath a smooth superficies, and if it coagulate, like a drop of tallow, and do not fall off the knife when it is turned downward, which also may be judged by its hanging like oyl to the sides of the Scummer. When the Liquor is brought to this pass, every hundred weight of it containeth about three-score and ten pounds weight of *Peter*.

When you find your Liquor thus ready to shoot, you must with great Iron Ladles lade it out of the Copper into a high narrow Tub for that purpose, which the Workmen call their settling Tub; and when the Liquor is grown so cold, that you can endure your finger in it, you shall find the common or cubick Salt begin to gravulate and stick to the sides of the Tub; then at the Tap, placed about half a foot from the bottom, draw off your Liquor into deep wooden Trays, or Brass-pans, and the cooler the place is where you let them stand to shoot in, the better and more plentifully will the *Salt-peter* be produc'd; but it will be of no good colour till it be refined, but will be part white, part yellow, and some part of it blackish.

The Salt which sticketh to the sides and bottom of the settling Tub is (as I have said) of the nature of common Salt; and there is scarce any *Peter* to be found but is accompanied with it, though no doubt some of this is drawn out of the Ashes by the second Liquors: If it be foul, they refine it by it self, and about London sell it at good rates to those that salt Neats Tongues, Bacon, and Collar-Beef; for besides a savory taste, it gives a pleasing red colour to most Flesh that is salted with it. Pliny says, *Nitrum obsonia alba et deteriora reddit Olera viridiora*, whether Salt-peter doth so, I have not yet tryed.

When the Liquor hath stood two Days and two Nights in the Pans, that part of the Liquor which is not coagulated but swims upon the *Peter*, must be carefully poured off, and being mingled with new Liquors, must again pass the Ashes before it be boiled, else it will grow so greasy it will never generate any Salt.

WILLIAM, VISCOUNT BROUNCKER:
from Thomas Sprat's *History of the Royal Society* (1667)

Experiments of the Recoiling of Guns[1]

WHEN I was commanded by this Society, to make some Experiments of the Recoiling of Guns: In order to the discovery of the cause thereof, I caused this Engine that lies here before you to be prepared, and with it (assisted by some of the most eminent of this Society) I had divers shots made in the Court of this College, near the length thereof from the mark, with a full charge (about a fourpenny weight) of Powder; but without any other success, than that there was nothing regular in that way, which was by laying it upon a heavy Table, unto which it was sometimes fastened with Screws at all the four places *R*, *L*, *V*, *B*, sometimes only at *R* or *L*, having wheels affixed at *L* and *V*, or *R* and *B*, that it might the more easily recoil.

This uncertainty I did then conceive might arise from one or more of these three causes, *viz.*

1. The violent trembling motion of the Gun, whence the Bullet might casually receive some literal impulse from the nose of the Piece at the parting from it.

2. The yielding of the Table, which was sensible.

3. The difficulty of aiming well by the Sight and Button so far from the Mark.

Therefore to avoid all these, the Experiments I caused to be made before you in the Gallery of this Colledge, you may be pleased to remember were performed, first, taking only eight grains of Powder for the charge. Secondly, laying the Engine upon the Floor; and, Thirdly, aiming by a Thread at *M*, a Mark* about an Inch and ¾ from the Mouth of the Gun (the edge of a knife being put for the Mark, the better to discern the line that was shot in) and they thus succeeded.

[1] The original paper contains a diagram.

When the Piece was fastned to the Floor both at *R* and *L*, the Bullet then did so fully hit the Mark, that it was divided by it into two parts, whose difference in weight was less than ten grains (about the thirty third part of the whole Bullet) although the lesser part was a little hollow, and that from which the neck of Lead was a little too close pared off: But when hindred from Recoiling only at *R*, the Bullet mist the mark towards *L* or *A*, for the whole Bullet, less than two grains excepted, went on that side: And in like manner when hindred from Recoiling at *L*, the Bullet mist the Mark towards *R* or *B*, the whole Bullet, less than two grains excepted, passing the knife on that side thereof.

I had the honour to make other Experiments with the same Engine, lately at White-Hall, before his Majesty and his Highness Royal within the Tilt-yard Gallery, where there is the hearth of a chimney raised a little above the Floor, about the distance of thirteen feet from the opposite wall, against which I caused a plank to be placed, and the Engine to be laid first against the middle of the Hearth, that it might not recoil at all, and that part of the board to be marked against which 'twas levelled, known by a line stretched from the Breech of the Piece unto the Board, directly over the sight and button; and the fire being given (the charge being but eight grains of Powder as before) the Bullet did fully hit the mark. Secondly, the Piece (charged and levelled in the same manner) was laid at the end of the Hearth next the Park, so that very little of the corner *R* rested against it, and then the Bullet miss'd the mark about an inch and a quarter towards the Park, or *A*. The like being done at the other end of the Hearth, the Bullet then miss'd the mark as much the other way; and afterwards with double that charge something more, as before I had found it less with a smaller charge.

Prince Rupert of Bavaria: from Robert Hooke's *Micrographia* (1665)

To make small shot of different sizes: Communicated by his Highness P.R.

Take Lead out of the Pig what quantity you please, melt it down, stir and clear it with an iron Ladle, gathering together the blackish parts that swim at top like scum, and when you see the colour of the clear Lead to be greenish, but no sooner, strew upon it *Auripigmentum* powdered according to the quantity of Lead, about as much as will lye upon a half Crown piece will serve for eighteen or twenty pound weight of some sorts of Lead; others will require more, or less, After the *Auripigmentum* is put in, stir the Lead well, and the *Auripigmentum* will flame: when the flame is over, take out some of the Lead in a Ladle having a lip or notch in the brim for convenient pouring out of the Lead, and being well warmed amongst the melted Lead, and with a stick make some single drops of Lead trickle out of the Ladle into water in a glass, which if they fall to be round and without tails, there is *Auripigmentum* enough put in, and the temper of the heat is right, otherwise put in more. Then lay two bars of Iron (or some more proper Iron-tool made on purpose) upon a Pail of water, and place upon them a round Plate of Copper, of the size and figure of an ordinary large Pewter or Silver Trencher, the hollow whereof is to be about three inches over, the bottom lower than the brims about half an inch, pierced with thirty, forty or more small holes; the smaller the holes are, the smaller the shot will be; and the brim is to be thicker than the bottom to conserve the heat the better.

The bottom of the Trencher being some four inches distant from the water in the pail, lay upon it some burning Coles to keep the Lead melted upon it. Then with the hot Ladle take

Lead off the pot where it stands melted, and pour it softly upon the burning Coles over the bottom of the Trencher, and it will immediately run through the holes into the water in small round drops. Thus pour on new Lead still as fast as it runs through the Trencher till all be done, blowing now and then the Coles with hand-bellows when the Lead in the Trencher cools so as to stop from running.

ROBERT HOOKE:
from *Micrographia* (1665)

A Prediction of Artificial Silk

AND I have often thought, that probably there might be a way found out, to make an artificial glutinous composition, much resembling, if not full as good, nay better, than that Excrement, or whatever other substance it be out of which, the silk-worm wire-draws his clew. If such a composition were found, it were certainly an easie matter to find very quick ways of drawing it out into small wires for use. I need not mention the use of such an Invention, nor the benefit that is likely to accrue to the finder, they being sufficiently obvious. This hint therefore, may, I hope, give some Ingenious inquisitive Person an occasion of making some trials, which if successfull, I have my aim, and I suppose he will have no occasion to be displeas'd.

TRANSLATIONS

THOMAS SALUSBURY:
from *The Systeme of the World in Four Dialogues*, by Galileo Galilei (tr. 1661)

Aristotle and the Telescope

Salviatus: There are certain Gentlemen yet living, and in health, who were present, when a Doctor, that was Professor in a famous Academy, hearing the description of the *Telescope*, by him not seen as then, said, that the invention was taken from *Aristotle*, and causing his works to be fetch't, he turned to a place where the Philosopher, gives the reason, whence it commeth, that from the bottom of a very deep Well, one may see the stars in Heaven, at noon day; and, addressing himself to the company, see here, saith he, the Well, which representeth the Tube, see here the gross vapours, from whence is taken the invention of the Crystals, and see here lastly the sight fortified by the passage of the rays through a diaphanous, but more dense and obscure *medium*.

Sagredus: This is a way to comprehend all things knowable, much like to that wherewith a piece of marble containeth in it, one, yea a thousand very beautiful Statua's, but the difficulty lieth in being able to discover them; or we may say, that is like to the prophecies of Abbot *Joachim*, or the answers of the Heathen *Oracles* which are not be understood, till after the things fore-told are come to passe.

Salviatus: And why do you not add the predictions of the *Genethliacks*, which are with like cleernesse seen after the event, in their Horoscopes, or, if you will, Configurations of the Heavens.

Sagredus: In this manner the Chymists find, being led by their melancholly humour, that all the sublimest wits of the world have writ of nothing else in reality, than of the way to make gold; but that they might transmit the secret to posterity without discovering it to the vulgar, they contrived some one way, and some another to conceal the same under several maskes; and it would make one merry to hear their comments upon the ancient *Poets*, finding out the important misteries, which lie hid under their Fables . . .

Simplicius: I believe, and in part know, that there want not in the world very extravagant heads, the vanities of whom ought not to redound to the prejudice of Aristotle, of whom my thinks you speak sometimes with too little respect, and the onely antiquity and bare name that he hath acquired in the opinions of so many famous men, should suffice to render him honourable with all that professe themselves learned.

Salviatus: You state not the matter rightly, *Simplicius*; there are some of his followers that fear before they are in danger, who give us occasion, or, to say better, would give us cause to esteem him lesse, should we consent to applaud their *Capricios*. And you, pray tell me, are you for your part so simple, as not to know that had *Aristotle* been present, to have heard the Doctor that would have made him Author of the *Telescope*, he would have been more displeased with him, than with those, who laught at the Doctor and his comments? Do you question whether *Aristotle*, had he but seen the novelties discovered in Heaven, would not have changed his opinion, amended his Books, and embraced the more sensible Doctrine; rejecting those silly Gulls, which too scrupulously go about to defend what ever he hath said. . . .

Philemon Holland:
from his translation of Pliny's *Natural History* (1601)

The Sundry Kinds of the Load-stone, and the Medicines thereto depending

Now that I am to passe from marbles to the singular & admirable natures of other stones; who doubts but the Magnet or Loadstone will present it self in the first place? for is there any thing more wonderfull, and wherein Nature hath more trauelled to shew her power than in it? True it is, that to rockes and stones she had giuen voice (as I haue already shewed) whereby they are able to answer a man, nay, they are ready to gainsay and multiply words vpon him. But is that all? what is there to our seeming more dull than the stiffe and hard stone? And yet behold, Nature hath bestowed vpon it, sence, yea &hands also, with the vse thereof. What can we deuise more stubborne and rebellious in the own kind, than the hard yron, yet it yeelds and will abide to be ordered: for loe, it is willing to be drawne by the load stone: a maruellous matter that this mettall, which tameth and conquereth all things els, should run toward I wot not what, and the nearer that it approcheth, standeth still as if it were arrested, and suffereth it selfe to be held therwith, nay, it claspeth and clungeth to it, and will not away. And hereupon it is, that some call the load-stone Sideritis, others Heracleos. As for the name Magnes that it hath, it tooke it (as Nicander saith) of the first inuentor and deuiser thereof who found it (by his saying) vpon the mountaine Ida (for now it is to be had in all other countries, like as in Spaine also;) and (by report) a neat-heard he was: who, as he kept his beasts vpon the foresaid mountaine, might perceiue as he went vp and downe, both the hob-nailes which were in his shooes, and also the yron picke or graine of his staffe, to sticke vnto the said stone. Moreover, Sotacus ascribeth and setteth downe fiue sundry kinds of the load-stone: the first which cometh out of Æthyopia; the second, from that Magnesia which confineth vpon Macedonie, and namely, on the right hand, as you go from thence toward the lake Bœbeis; the third is found in Echium, a town of Bœotia; the fourth about Alexandria, in the region of Troas; and the fift in Magnesia, a country in Asia Minor. The principall difference obserued in these stones, consists in the sex (for some be male, others female;) the next lieth in the colour. As for those which are brought out of Macedonie and Magnesia, they be partly red, and partly blacke. The Bœotian loadstone standeth more vpon red than black: contrariwise, that of Troas is black, and of the female sex, in which regard it is not of that vertue that others be. But the worst of all comes from Magnesia in Natolia, and the same is white: neither doth it draw yron as the rest, but resembles the pumish stone. In sum, this is found by experience, That the blewer any of these loadstones be, the better they are and more powerful. And the Ethyopian is simply the best, insomuch, as it is worth the weight in siluer: found it is in Zimiri, for so they cal the sandy region of Ethyopia, which country yeeldeth also the sanguine load-stone, called Hæmatites, which both in color resembleth bloud, and also if it be bruised, yeeldeth a bloody humour, yea and otherwhiles that which is like to saffron. As for the property of drawing yron, this bloud-stone Hæmatides is nothing like to the loadstone indeed. But if you would know and try the true Ethyopian Magnet, it is of power to draw to it any of the other sorts of loadstones. This is a generall vertue in them all, more or lesse, according to that portion of strength which Nature hath indued them withal, That they are very good to put into those medicines which are prepared for the eies: but principally they do represse the vehement flux of humors that fall into them: beeing calcined and beaten into pouder, they do heale any burne or scald. To conclude, there is another mountaine in the same Ehyopia, and not far from the said Zimiris, which breedeth the stone Theamedes that will abide no yron, but rejecteth and driueth the same from it. But of both these natures, as well the one as the other, I haue written oftentimes already.

The first inuention of glasse, and the manner of making it. Of a kind of Glasse, called obsidianum. Also of sundry kindes of Glasse, and those of many formes.

There is one part of Syria called Phœnice, bordering upon Iurie, which at the foot of the mount Carmell, hath a meere named Cendeuia; out of which the river Belus is thought to spring, and within fiue miles space, falleth into the sea, near vnto the colony Ptolemais: This river runneth but slowly and seemeth a dead or dormant water, vnwholesome for drinke, howbeit, vsed in many sacred ceremonies with great deuotion; full of mud it is, and the same very deepe ere a man shall meet with the firm ground: and vnlesse it be at some spring tide, when the sea floweth vp high into the riuer, it neuer sheweth sand in the bottom; but then, by occasion of the surging waues, which not only stir the water, but also cast vp & scoure away the grosse mud, the sand is rolled too and fro, and being cast vp, sheweth very bright and cleare, as if it were purified by the waues of the sea: and in truth, men hold opinion, That by the mordacity and astringent quality of the salt water, the sands become good, which before serued to no purpose. The coast along this riuer which sheweth this kind of sand, is not aboue half a mile in all, and yet for many a hundred yeare it hath furnished all places with matter sufficient to make glasse. As touching which deuise, the common voice and fame runneth, that there arriued sometimes certain merchants in a ship laden with nitre, in the mouth of this riuer, & being landed, minded to seeth their victuals vpon the shore and the very sands: but for that they wanted other stones, to serue as treuets to beare vp their pans and cauldrons ouer the fire, they made shift with certain pieces of sal nitre out of the ship, to support the said pans, and so made fire vnderneath: which being once afire among the sand and grauell of the shore, they might perceiue a certaine cleare liquor run from vnder the fire in very streams, and hereupon they say came the first inuention of making glasse. But afterwards (as mans wit is very inuentiue) men were not content to mix nitre with this sand, but began to put the Load-stone among, for that it is thought naturally to draw the liquor of glasse vnto it, as well as yron. Then they fell to calcine and burne in many other places shining grauell stones, shels of fishes, yea, and sand digged out of the ground, for to make glasse therewith. Moreouer, diuers authors there be who affirme, That the Indians vse to make glasse of the broken pieces of Crystall, and therefore no glasse comparable to that of India. Now the matter whereof glasse is made, must be boiled or burnt with a fire of dry wood, and the same burning light and cleare without smoke, and there would be put thereto brasse of Cypros, and nitre, especially that which commeth from Ophyr. The furnace must bee kept with fire continually, after the manner as they vse in melting the ore of brasse. Now the first burning yeeldeth certaine lumps of a fatty substance, and blackish of colour. This matter is so keen and penetrant whiles it is hot, that if it touch or breath vpon any part of the body, it will pierce and cut to the very bone ere one be aware or do feele it. These masses or lumps be put into the fire againe, and melted a second time in the glasse houses, where the colour is giuen that they shall haue: and then some of it with blast of the mouth, is fashioned to what form or shape the workman will: other parcells polished with the Turners instrument, and some againe engrauen, chased, and embossed in manner of siluer plates: in all which feats, the Sidonians in times past were famous artificers: for at Sidon were deuised also mirroirs or looking glasses. Thus much as touching the antique maner of making glasse. But now adaies there is a glasse made in Italy of a certain white sand, found in the riuer Vulturnus for six miles space along the shore towns, from the mouth where he dischargeth himselfe into the sea, and this is between Cumes and the lake Lucrinus. This sand is passing soft and tender, whereby it may be reduced very easily into fine pouder, either to be beaten in morter or ground in mill: to which pouder the manner is to put three parts of nitre, either in weight or measure, and after it is the first time melted, they vse to let it passe into other furnaces, where it is reduced into a certain masse, which, because it is compounded of sand and nitre, they call Ammonitrum: this must be melted againe, and then it becomes pure glasse, and the very matter indeed of the white clear glasses: and in this sort throughout France and Spain the maner is to temper their sand, and to prepare it for the making of glasse. Moreouer, it is said, That during the reigne of Tiberius the Emperor, there was deuised a certain temper of glasse, which made it pliable and flexible to wind and turne without breaking: but the artificer who deuised this, was put downe, and his work-house, for feare lest vessels made of such glasse should take away the credit for the rich plate of brasse, siluer, and gold, and make them of no price: and verily, this bruit hath run currant a long time (but how true, it is not so certain.) But what booted the abolishing of glasse-makers, seeing that in the daies of the Emperor Nero the art was growne to such perfection, that two drinking cups of glasse (and those not big, which they called Pterotos) were sold for 6000 sesterces.

There may be ranged among the kinds of glasses, those which they call Obsidiana, for that they carry some resemblance of that stone, which one Obsidius found in Æthyopia; exceeding blacke in colour, otherwhiles also transparent: howbeit, the sight therein is but thicke and duskish. It serueth for a mirroir to stand in a wall, and instead of the image yeeldeth back shadows. Of this kind of glasse many haue made jewels in maner of precious stones: and I my selfe haue seene massiue pourtraitures made thereof, resembling Augustus late Emperor of famous memory, who was wont to take pleasure in the thicknes of this stone, insomuch as he dedicated in the temple of Concord for a strange and miraculous matter, foure Elephants made of this Obsidian stone. Also Tiberius Cæsar sent back again to the citizens of Heliopolis, a certain image of prince Menelaus, found among the moueable goods of one who had bin lord gouernor in Ægypt, which he had taken away out of a temple, among other cerimoniall reliques: and the said statue was all of the Iaiet, called Obsidianus. And by this it may appeare, That this matter began long time before to be in vse, which now seemeth to be renued again and counterfeited by glasse that resembleth it so neare. As for the said Obsidian stone, Xenocrates writeth, That it is found naturally growing among the Indians; within Samnium also in Italy, and in Spaine along the coast of the Ocean. Moreouer, there is a kind of Obsidian glasse, with a tincture artificiall, as black as Iaiet, which serueth for dishes and platters to hold meat: like as other glasse, red throughout, and not transparent, called for that colour Hæmatinon. By art likewise there be vessels of glasse made white and of the colour of Cassidony, resembling also the Iacinct and Saphire, yea & any other colors whatsoeuer. In sum, there is not any matter at this day more tractable and willing either to receiue any forme or take a color, than glasse: but of all glasses, those be most in request and commended aboue the rest, which be white, transparent and cleare throughout, comming as neare as it is possible to Crystall. And verily, such pleasure do men take now adaies in drinking out of faire glasses, that they haue in maner put downe our cups and boules of siluer or gold: but this I must tell you, that this ware may not abide the heat of the fire, vnlesse some cold liquor were put therin before: and indeed, hold a round bal or hollow apple of glasse ful of water against the Sun, it will be so hot, that it is ready to burne any cloth, that it toucheth. As for broken glasses, well may they be glued and sodered againe by a warme heat of the fire, but melted or cast again they cannot be whole, vnlesse a man make a new furnace of pieces broken from another: like as we see there be made counting rundles thereof, which some call Abaculos, whereof some are of diuers and sundry colors. Moreouer, this would be noted, That if glasse and sulphur be melted together, they will souder and vnite into a hard stone. To conclude, hauing thus discoursed of all things that are knowne to be done by wit or art, according to the direction of Nature, I cannot chuse but maruell at fire and the operations thereof, seeing that nothing in a manner is brought to perfection but by fire; and thereby any thing may be done.

NOTES

p. 43. **Francis Bacon: "Description of Salomon's House"**

p. 43. *Francis Bacon* (1561–1626). Lawyer, statesman, philosopher, and author. Baron Verulam, Viscount St Albans, and Lord Chancellor. "The wisest, greatest, meanest of mankind." Advocated the true method of progress in science by experimental research. *The New Atlantis* was written before 1617 and published in 1627. It describes an ideal country whose inhabitants are devoted to scientific investigations.

p. 43. *Salomon's House.* The scientific research institute, with laboratories, in the imaginary country called the New Atlantis.
p. 43. *coagulations.* Solidifications.
p. 43. *indurations.* Hardenings, temperings.
p. 43. *conservations.* Preserving, keeping.
p. 44. *perspective houses.* Optical laboratories.
p. 44. *vitrificated.* Vitrified.
p. 45. *sound houses.* Acoustic laboratories.
p. 45. *quarter sounds.* Quarters of a musical interval of one major tone. A quarter tone is one half of a semitone.
p. 45. *lesser slides of sounds.* Lesser acoustic intervals.
p. 45. *great sounds extenuate.* Long drawn out and loud sounds.
p. 45. *trunks.* Pipes, tubes.
p. 45. *sallets.* Salads.
p. 45. *engine houses.* Workshops.
p. 45. *basilisks.* Large brass cannon firing shot of about two hundred pounds.
p. 46. *brooking of seas.* Standing up to rough seas.
p. 46. *motions.* Machines, contrivances, apparatus.
p. 46. *swimming girdles and supporters.* Life belts and water-wings.
p. 46. *motions of return.* Machines performing oscillatory motion.
p. 47. *pioners.* Pioneers.

p. 48. **Thomas Sprat: "The History of the Royal Society"**

p. 48. *Thomas Sprat* (1635–1713). Bishop of Rochester. In 1667, after the Royal Society had only existed five years from the granting of its charter, Sprat wrote *The History of the Royal Society*, a kind of interim report on its achievements, and a prospectus of future work. It was a reply to critics of the experimental method of advancing science.

The Work of Christopher Wren

p. 48. *Dr Christopher Wren.* Later Sir Christopher Wren (1632–1723), the famous architect of St Paul's and also a very able scientist.
p. 48. *gross Trials.* Rough tests, pilot experiments.
p. 48. *globous.* Spherical.
p. 49. *Rundle.* Drum, cylinder.
p. 50. *a Natural standard for Measure from the Pendulum for vulgar use.* Samuel Butler seems to be hitting at this statement in *Hudibras*, Part II, Chapter 2, verse 1022:

> Upon the Bench I will so handle 'em
> That the Vibration of this Pendulum
> Shall make all Taylor's Yards of one
> Unanimous Opinion:
> A Thing he long has vapour'd of
> But now shall make it out by proof.

p. 50. *luxation of the Instrument.* Dislocation or straining of the instrument.
p. 50. *Retes.* Spider lines or cross wires in the eyepiece of the telescope.
p. 51. *Crepusculine Observations.* Observations made in twilight.
p. 51. *Hugonius.* Christian Huyghens (1629–95), a famous Dutch scientist and astronomer.

Prospects of Scientific Discovery in America and Elsewhere

p. 52. *Virtues.* Properties.
p. 53. *The Invention of Longitude.* A method of finding accurately the longitude of ships at sea. The compensated chronometers invented by John Harrison (1693–1776) went far to solve this problem.

Mechanical Inventions

p. 56. *some lucky Tide of Civility.* Some lucky influx of civilization.
p. 57. *Graving.* Carving, sculpture.

p. 57. *Limning.* Painting.
p. 57. *Powder.* Gunpowder, the European invention of which is sometimes ascribed to Berthold Schwartz, a German monk, about the fourteenth century.
p. 57. *Bow-dye.* A scarlet dye, named after Bow, or Stratford-le-Bow, in East London.
p. 57. *Art of Composing Letters.* The art of printing.
p. 59. *contemn'd.* Despised, treated with contempt. Has the same meaning in the authorized version of the Bible.

p. 59. **Isaac Newton: " Letter to Henry Oldenburg "**
p. 59. *Isaac Newton* (1642–1727). Perhaps the greatest scientist that ever lived. Discovered the law of universal gravitation, the differential calculus, and the dispersion of light. Knighted in 1705.

p. 59. *Mr Oldenburg.* Henry Oldenburg (1615–77) was secretary of the Royal Society.
p. 59. *Mr Linus.* Probably Francis Line, alias Hall (1595–1673), a Jesuit and scientific writer. Entered controversies with Newton and Boyle. Famous as a maker of sun-dials. See D.N.B.
p. 60. *a fermental principle.* A process of fermentation.
p. 60. *contextures.* Assemblages, mixtures.
p. 60. *the Protoplast.* God as Creator of first things.
p. 60. *inspissated.* Coagulated.

p. 62. **John Ray: " Of the Number of Plants "**
p. 62. *John Ray* (1627–1705). A very eminent botanist and naturalist. Collaborated with his friend Francis Willoughby in writing books on these subjects. The *Philosophic Letters* were published by W. Derham in 1718. See the article in the *Dictionary of National Biography* for a fuller account of Ray.

p. 62. *anon.* Immediately, very soon. *Cf.* the waiter, Francis, in *Henry IV, Part I*, in the tavern scene. "Anon, anon, sir!"
p. 64. *Pasterns.* The pastern is that part of a horse's leg between the hoof and the fetlock. The fetlock is that part of a horse's leg, a short distance above the hoof, where a tuft of hair grows. The famous Dr Samuel Johnson, in the first edition of his *Dictionary*, defined the pastern as the knee of a horse. To a lady who inquired how he came to make such a mistake, he replied "Ignorance, madam! Pure ignorance!"
p. 64. *Haselbedge in the Peak of Derbyshire.* Hazelbadge Hall lies

between the villages of Bradwell and Little Hucklow, not far from Castleton.

p. 64. Nehemiah Grew: " An Idea of a Philosophical History of Plants "

p. 64. *Nehemiah Grew* (1641–1712), a famous botanist. Wrote *The Anatomy of Vegetables Begun* (1672), *An Idea of a Phytological History Propounded* (1673), *The Comparative Anatomy of Trunks* (1675), and *The Anatomy of Plants* (1682).

p. 64. *Vegetables.* Plants in general.

p. 65. *Clusius.* Jules Charles de l'Écluse (1526–1609), a botanist, born at Arras, Flanders. He published various works, of which the chief was *Rariorum Plantarum Historia* (1601).

p. 65. *Columna.* Fabio Colonna (1567–1650), usually called Fabius Columna, a botanist and lawyer, born at Naples. In 1592 he published a herbal called *Phytobasanos*, and in 1606 and 1616, two volumes of another called *Ekphrasis.*

p. 65. *Bauhinus.* This might be Gaspard Bauhin (1560–1624), a botanist, born at Basle. He published *Phytopinax Theatri Botanici* (1596), *Prodromus Theatri Botanici* (1620), and *Pinax Theatri Botanici* (1623). Equally probably it might be his brother, Jean Bauhin (1541–1613), also born at Basle. He wrote *Historia Universalis Plantarum Nova et Absolutissima*, of which a sketch was published in 1619, and a full edition in 1650–51.

p. 65. *Boccone.* Paolo Sylvia Boccone (1633–1703), an Italian botanist. Became a monk under the name of Sylvius. He published *Icones et Descriptiones Variarum Plantarum* (1674).

p. 65. *Mr Ray.* John Ray (1627–1705), the leading British naturalist of his time. For his work see p. 23.

p. 65. *Dr Morrison.* Robert Morison (1620–83), the first Professor of Botany at Oxford. For his work see p. 23.

p. 65. *Mr Evelyn.* John Evelyn (1620–1706), English virtuoso and diarist. Published *Sylva: or a discourse of Forest Trees* (1664).

p. 65. *Dr Beal.* John Beale (1603–83), English scientific writer. Published *Aphorisms concerning Cider* (1664) and *Herefordshire Orchards* (1656), and various papers in the *Philosophical Transactions.* A friend of John Evelyn, the diarist.

p. 65. *Goat's Rue.* Probably the plant still called by that name or French Lilac. "A stout, erect, hairy perennial up to four feet in height, with white lilac or pink flowers."

p. 66. *Centaurium.* Common Centaury, is called *Centaurium minus* under the 1952 classification. There appears to be now no *Centaurium majus*, but Common Knapweed has been popularly called Greater Centaury.

p. 66. *Chelidonium.* Greater Celandine is still called *Chelidonium majus*. There is now no *Chelidonium minus*, but the Lesser Celandine once had that name.

p. 66. *Affections.* Properties, qualities, or atttributes.

p. 68. *Teeming.* The act of seeding or fruition.

p. 68. *Œconomy.* Arrangement of the conditions of living.

p. 68. *officious.* Helpful, assisting.

p. 69. **Martin Lister: "The Nature and Difference of the Juices of Plants"**

p. 69. *Martin Lister* (*c.* 1638–1712), physician and zoologist. Correspondent of John Ray. Wrote *Historia sive Synopsis Methodica Conchylorum* (1685–92) and other works.

p. 69. *exsudate.* Exude, emerge through the pores. Literally 'sweat out.'

p. 69. *Blebs.* Swellings, pustules.

p. 69. *the Hypericum Kind.* The St John's Wort tribe.

p. 69. *Rorella.* Round-leaved Sundew.

p. 69. *Androsæmum Hypericoides Ger.* The plant of the St John's Wort tribe called Common Tutsan. "Ger." means "in the *History of Plants*, by John Gerard (1633)." Might possibly be "Hairy St John's Wort."

p. 70. *Hypericon Ascyron Dictum, Caule Quadrangulo J.B.* Quadrangular St John's Wort, also called St Peter's Wort.

p. 70. *Hypericum pulchrum tragi, J.B.* Upright St John's Wort. J.B. probably means "so-called in Thomas Johnson's *Mercurius et Itinera Botanica* (1634)."

p. 70. *diverse Faculties.* Various organs, or various characteristics.

p. 70. *V.C.P.A.* I have not yet been able to interpret these letters.

p. 70. *Milk.* Juice, sap, latex.

p. 70. *Spondylium Ger.* Common Cow-parsnip.

p. 70. *Lactuca syl. costa spinosa, C.B.* Prickly Lettuce. C.B. means "so-called in *Centuria Plantarum Gedani*, by James Breyn (1678)."

p. 70. *ropes.* Forms ropes, strings, or fibres.

p. 70. *in the Shell I drew it.* In the shell in which I collected it.

p. 70. *the Caseous Part.* Having the nature of cheese. The curds.

p. 71. *Cags.* Stiff points, stumps, little hard lumps.

p. 71. *Menstruum.* Solvent.

p. 71. *S.V. Spiritus Vini.* Methylated spirits.
p. 71. *A Day's Insolation.* Standing for a day in the sun.
p. 71. *the Trachelium kind.* Bell Flowers, Canterbury Bells.
p. 71. *Tithymallus Helioscopius Ger.* Sun Spurge or Water Wort.
p. 71. *Flower.* Flour.
p. 71. *Essence-Bottles.* Scent bottles.
p. 72. *let go.* Separate from, part with.
p. 72. *Sonchus, lævis and asper.* Common Sow Thistle, smooth and rough varieties. *Lævis* should be *levis*, smooth.
p. 72. *Papaver Rheas Ger.* Common Red Poppy.
p. 72. *it roaped.* It formed threads, strings or ropes.
p. 72. *clammy.* Viscous.
p. 72. *Trogopogon flore luteo J.B. Tragopogon* is the Latin name of Goat's Beard, both yellow and purple.
p. 72. *Convolvulus major J.B.* Common Bindweed or Convolvulus.
p. 72. *Chelidonium majus Ger.* Greater Celandine.
p. 73. *Centaurium luteum perfoliatum C.B.* One of the Centauries.
p. 73. *Angelica sativa Park.* The plant Angelica. Park means "so-called in the *Theatrum Botanicum of John Parkinson* (1640)."
p. 73. *Fountain water.* Spring water.
p. 73. *serous.* Waxy.
p. 74. *Pericylmenum Ger.* Common Honeysuckle or Woodbine.
p. 74. *Pinguicula.* Common Butterwort or Yorkshire Sanicle.
p. 74. *the Chats of the Alder.* Catkins of the Alder.
p. 74. *Lapatha.* The Dock family. A name also given to Goose-foot or English Mercury.
p. 75. *desecated Juices.* Desiccated, dried juices.
p. 75. *Mr Fisher.* I have not yet identified this observer. He is not named in Pepys's *Diary*, Evelyn's *Diary*, or *Philosophical Transactions*, and Name Lists of the Royal Society, or in Gunther's *Early British Botanists.* There is no article on him in the *Dictionary of National Biography.*
p. 75. *Acetosa.* This name is applied to (1) Mountain Dock, (2) Common Dock, and (3) Sheep's Dock.
p. 75. *Languedoc.* A large province in the south-east of Old France, including the lower Rhône valley, Toulouse, the Cévennes, the Mediterranean coastal belt, the south-east part of the Massif Central, and the Causses.
p. 75. *sour Docken.* Docken is still often used as the plural of the plant dock.
p. 76. *Infusion or Maceration of Rhubarb.* Infusion is "the solvent action of boiling water on vegetable drugs during the time

occupied in cooling." Maceration "consists in subjecting a mixture of soluble and insoluble matter (*i.e.*, the rhubarb) to the solvent action of fluids (*e.g.*, water) at ordinary temperatures"—*i.e.*, in steeping the rhubarb in water.

p. 76. *Lauro-cerasus. Cerasus* is a cherry-tree or cherry. The name *Lauro-cerasus* seems to mean laurel-cherry, but the very full list of plants and trees in W. Hudson's *Flora Anglica* (1799) does not contain this.

p. 76. *Lime.* A sticky substance like glue. *Cf.* bird lime.

p. 77. *S.S.S. Stratum super stratum.* In layers.

p. 77. *dissected.* Cut open.

p. 77. *Helenium, sive Enula Campana J.B.* The plant called Elecampane, akin to the Aster.

p. 77. *Cicuta.* Water hemlock or Cowbane.

p. 78. *Tapsus barbatus Ger. Thapsus Barbatus* was Great Mullein or High Taper.

p. 78. *Baccæ Lauri.* Laurel berries.

p. 78. *Hederæ.* Common Ivy.

p. 78. *Cornus Fœminæ.* Wild Cornel.

p. 78. *Helleborus niger syl. adulterinus etiam Hyeme virens J.B.* Green Hellebore or Bear's Foot.

p. 78. *boaken.* Throb, ache.

p. 78. *Juniperus vulgaris, baccis parvis purpureis J.B.* Common Juniper, with small purple berries.

p. 78. *the Chops of Ivy made in March.* Chips, splinters of Ivy.

p. 79. *Lactuca syl. Costa spinosa C.B.* Prickly Lettuce.

p. 79. *the blue flower of ripe Plums.* The 'bloom' on the surface of the plums.

p. 79. *Bonus Henricus J.B.* The plant Good King Henry.

p. 79. *called unctuous by C.B.* Called oily in *Centuria Plantarum Gedani*, by James Breyn (1678).

p. 79. *Glicyrrhiz. Astragalus*, or Wild Liquorice.

p .80. *the Wicking.* Probably the Mountain Ash or Rowan Tree. A common form of the name is 'Quicken.'

p. 80. *to depurate or desecate the Opium.* That is, to purify or dry the opium.

p. 80. *Lac.* Resin or gum, usually from the East Indies.

p. 80. *Benzoin.* An aromatic resinous gum.

p. 80. *Jallap* (now Jalap). A purging medicine, called after Xalapa in Mexico.

p. 80. *Guiacum.* The resin of certain evergreen trees, *Guaiacum officinale* or *Guaiacum sanctum*, natives of the West Indies. The resin is a mild purgative.

p. 80. *Lact. syl.* The Prickly Lettuce mentioned earlier.

p. 80. Martin Lister, John Ray, and Tancred Robinson: "Fungi"

p. 80. *Martin Lister* (*c.* 1638–1712). See p. 193.

p. 80. *John Ray* (1627–1705). See p. 191.

p. 80. *Tancred Robinson* (*c.* 1660–1748). Physician and naturalist. Correspondent of John Ray. Knighted by George I, to whom he was physician in ordinary. Wrote papers in the *Philosophical Transactions* on mineralogy, geology, botany, and meteorology.

p. 80. *Marton Woods under Pinno Moor in Craven.* Pinhaw, or Pinhow, is a hill nearly 1300 feet high, two to three miles south-west of Skipton-in-Craven, Yorkshire. East and West Marton are two villages a few miles to the north-west of Pinhaw. Martin Lister, though born elsewhere, was a member of a county family living in the district. He had property there. He was related to the wife of General John Lambert.

p. 81. *Jon Bauhin.* Probably Jean Bauhin (1541–1613), a native of Basle. A botanist and physician. He compiled a vast encyclopædia called *Historia Universalis Plantarum Nova et Absolutissima.*

p. 81. *Fungus porosus crassus magnus J.B.* J.B. means "so-called in Thomas Johnson's *Mercurius et Itinera Botanica* (1634)."

p. 81. *will . . . fix in a Purple.* Will end up by assuming a purple colour.

p. 81. *a chance Original.* An accidental origin.

p. 82. *Alcea's.* Probably the plant now called Vervain Mallow, which formerly had the Latin names *Alcea vulgaris* and *Malva verbenacea.* Note the peculiar form of the plural.

p. 83. *a hircine Odour.* A goatish smell.

p. 83. *Arachidna's.* The now well-known pea-nuts or ground-nuts, which in our school-days we called monkey-nuts.

p. 83. *non nisi alta Fossione inveniendæ.* Not found without deep digging.

p. 84. Edmund King, Francis Willoughby, and Martin Lister: "The Generation of a Sort of Bees in old Willows"

p. 84. *Edmund King* (1629–1709), physician to Charles II, whom he attended on his deathbed. Knighted 1676. Wrote papers in the *Philosophical Transactions* on insects, animalculæ, and transfusions of blood. Made skilful dissections.

p. 84. *Francis Willoughby or Willughby* (1635–72), naturalist. Friend collaborator, and patron of John Ray. wrote on the natural history of birds, beasts, fishes, and insects.

p. 84. *Sir J. Bernhard.* A Northamptonshire knight of no great importance.
p. 84. *close.* Air-tight.
p. 84. *cartrages of powder.* Cartridges, neat packages.
p. 84. *rabbets.* This word, with one vowel change, has replaced 'conies,' formerly in general use.
p. 85. *the Wells at Astrop.* A village in south-west Northamptonshire, with a spa well, which became known in 1660–70. The name was once Eastthorpe.
p. 85. *Mr Wray.* John Ray (1627–1705), the famous naturalist.
p. 86. *Nymphæ.* Nymphs. The young stages of insects which do not undergo metamorphosis during their growth. The plural is spelt Nympha's on p. 87.
p. 87. *blue pipe or syringe-tree.* The lilac. The mock orange or syringa was called the white pipe.
p. 87. *the Dam.* The mother insect.

p. 88. **John Winthrop, Henry Oldenburg, and [?] Hammersly: "Humming Birds"**

p. 88. *John Winthrop* (1606–76), known as John Winthrop the younger. Born in Suffolk. Served at the Isle of Rhé (1628). Migrated to New England. Governor of Connecticut (1660–76). F.R.S. 1662. Wrote papers in the *Philosophical Transactions* on (1) Some Natural Curiosities of New England, and (2) Maize.
p. 88. *Henry Oldenburg* (*c.* 1615–77), a Secretary of the Royal Society. A correspondent of Isaac Newton.
p. 88. [?] *Hammersly.* The Christian name of Mr Hammersly is not given. We are simply told that he was "of Coventry."

p. 88. *five grains.* Nearly a third of a gram, or, to be more exact, 0·3240 gram.
p. 89. *our English Drake's Heads.* Probably the head of the mallard drake, with its beautiful green.
p. 89. *a tit-mouse . . . weighed above two Shillings.* A newly minted coin has often been used as a standard weight. Also pieces of brass, of the same weight as new golden guineas and half-guineas, have been used to check that coins have not been clipped.
p. 89. *the relation that he is a curious singing bird, I think it untrue.* "I believe the assertion that the humming bird is a good singer to be untrue."

p. 89. *An Indian Soggamore.* Now more usually spelt Sagamore. A great chief, *e.g.*, Chingachgook in the Fenimore Cooper novels, or Pontiac in genuine American history.

p. 90. Robert Murray (or Moray), Tancred Robinson, and John Ray: " Barnacles "

p. 90. *Robert Murray* or *Moray* (?–1673). A founder of the Royal Society. Knighted by Charles I in 1643. Lord of the Exchequer for Scotland. Friend of Samuel Pepys. Soldier, courtier, and scientist.

p. 90. *Tancred Robinson* and *John Ray.* See the notes about these men on p. 196 and p. 191, respectively.

p. 90. *The island of East.* Probably Uist in the Outer Hebrides.

p. 90. *the little bird within it.* An allusion to the old notion that barnacle shellfish eventually fell into the sea and became barnacle geese.

p. 91. *the Sein.* The river Seine in Normandy.

p. 91. *the Addesis.* The river Adige in north-east Italy.

p. 91. *the Mare Mortuum.* Probably Mare Morto, the westerly of the two harbours at Cape Miseno, about fifteen miles west-south-west of Naples. Constructed by Agrippa in the time of Augustus.

p. 91. *the Lake Avernus.* A small deep freshwater lake about four miles north of Mare Morto. Famous in classical times as an entrance to the infernal regions. The name is often translated as "Avernus hole." The lake is an extinct volcanic crater filled up by water.

p. 92. *the cases.* Those parts of the bodies of the ducks left behind when the internal organs had been removed.

p. 92. *Tridactylæ.* A class of seagulls supposed to have three toes. The Latin name of the Kittiwake gull is *Rissa Tridactyla.* The bird really has four toes, although one is very small.

p. 92. *Mergi.* The class of sea birds now called Divers.

p. 92. *a Reservatory of Air.* A reservoir of air.

p. 92. *the plash-duck.* Perhaps the teal.

p. 94. John Woodward: "An Essay toward a Natural History of the Earth and Terrestrial Bodies"

p. 94. *John Woodward* (1661–1727). Physician. One of the founders of British geology. Recognized fossils as "real spoils of once living animals." Published the *Essay* in 1695, and *An Attempt towards a Natural History of the Fossils of England* in 1728–29.

The Nature of Fossils

p. 94. *Belemnites*. The fossil remains (guards) of certain cuttlefish of Mesozoic times.

p. 94. *Selenites*. Selenite is calcium sulphate (gypsum) in a certain crystalline or foliated form.

p. 94. *Marchasites*. Forms of the crystalline mineral, iron sulphide, with a golden lustre.

p. 94. *reposed*. Deposited.

p. 95. *Papillæ*. *Papulæ* are small tube-like projections on the skin of the starfish, forming its respiratory organs.

p. 95. *Sutures*. Lines on the surfaces of fossil shells, formed by the meeting of internal partitions called *septa* with the surface of the shell.

p. 95. *Balani*. Probably shellfish known as Acorn-Shells, akin to barnacles. For diagrams see *A Manual of Palæontology*, by Nicholson and Lydeker, p. 496.

p. 95. *Tubuli vermiculares*. Fossils of tubular form.

p. 95. *vires*. Strength or force.

p. 96. *Exuviæ*. Remains of dead animals. Fossils are therefore examples of Exuviæ.

"Concerning the Universal Deluge. That these Marine Bodies were then left at Land. The Effects it had upon the Earth"

p. 96. *at Land*. On land.

p. 96. *Consectaries*. Deductions.

p. 96. *Undertakers*. Investigators.

p. 97. *conferring*. Deciding, settling.

p. 98. *wave*. Waive, omit.

p. 98. *down to the Abyss*. The abyss supposed to lie under the firm crust of the earth.

p. 98. *assumed up*. Carried away at random and held in suspension by the water.

p. 99. *Echine*. Sea urchins, small sea animals with rough, prickly skins.

p. 100. **Richard Buckley (or Bulkeley), Samuel Foley, and Thomas Molyneux: " The Giant's Causeway in Ireland "**

The original is illustrated by four diagrams on one sheet. These three men were Fellows of the Dublin Philosophical Society.

p. 100. *Sir R. Buckley*. Sir Richard Bulkeley, Bart (1644–1710), an Irishman. Contributed a paper on *A New Sort of Calash* (1685), another on Maize (1693), and a third on Elmseed (1693) to the *Philosophical Transactions of the Royal Society*.

p. 100. *Dr Samuel Foley*. Samuel Foley (1655–95), Bishop of Down and Connor.

p. 100. *Dr Thomas Molyneux*. Thomas Molyneux (1661–1733), born at Dublin. Physician, F.R.S. (1687), created a baronet (1730). Contributed several papers on biology, geology, and medicine to the *Philosophical Transactions of the Royal Society*.

p. 100. *the Bishop of Derry*. In 1692 Dr William King (1650–1729), afterwards (1703) Archbishop of Dublin, was the Bishop of Derry, *i.e.*, of Londonderry, in N. Ireland.

p. 100. *Colrain*. Now Coleraine.

p. 102. *the Concave and the Convex Superficies*. The concave and convex surfaces.

p. 102. *the Contexture*. The structures, the shapes.

p. 102. *Graving*. Incisions, carving.

p. 103. *the Entrochos*. The wheel-like joints of the fossils called encrinites, found in carboniferous limestone.

p. 103. *the Astroites* or *Lapis Stellaris*. *Astrædiæ*, a family of star corals.

p. 103. *the Lapis Basanus* or *Basaltes*. *Lapis Lydius Basanus* is Lydian stone, or touchstone, a kind of slate on which gold and silver were rubbed, the colour of the streak showing the purity.

p. 105. *the Lapis Basaltes Misneus*. Now called simply basalt. An igneous rock forming columns or pillars, as at the Giant's Causeway.

p. 105. *Kentmannus in "Gesner de Figuris Lapidum."* Perhaps Johann Kentmann of Torgau, a collector of fossils in the seventeenth century. Conrad Gesner of Zurich (1516–65), the first botanist and zoologist of this time, studied fossils. His great work was *De Rerum Fossilium Lapidum et Gemmarum maxime figuris et similitudinibus Liber* (1565).

p. 105. *Lapides Angulosi . . . exploratum est*. Basalt has the form of a number of angular rocks, joined together, each having the shape and size of an ordinary fig. But there are so many individual stones that they are joined and fitted together as if by the skill of the cabinet-maker. The number of angles may be seven, six, or five, and sometimes, though more rarely, four. In general the shape of each stone is that of a vertical pillar, with a smooth surface, of the colour of iron, of high density, and as hard as steel. These stones, thus joined together, protrude seven to ten yards out of the ground. How deep they go, no one has yet discovered.

p. 105. *Lapis Basaltes . . . Geniculatus.* The Basalt Rock or Great Irish *Basanos* has at least three angles, at most eight. Each column consists of several segments, with exactly fitting joints, but easily separable. It may bend abruptly at an angle.

p. 106. **John Wallis: "To Find the Parallax of the Fixed Stars"**

p. 106. *Dr John Wallis.* Dr John Wallis (1616–1703), a famous mathematician and scholar.

p. 106. *Parallax of the Fixed Stars.* The angle subtended at a star by a radius of the earth's orbit is called the annual parallax of the star.

p. 106. *Systema Cosmicum.* Systems of the Universe. These are the first two words of the title of a Latin edition of Galileo's famous *Dialogi* (see note on p. 214), the first edition printed in England in 1663.

p. 106. *Dr Hook.* Robert Hooke, F.R.S. (1635–1703). See p. 202.

p. 106. *Mr Flamsteed.* John Flamsteed, F.R.S. (1646–1719), the first Astronomer Royal.

p. 106. *our Zenith.* The point of the sky vertically overhead.

p. 106. *the Zodiack.* A belt of the celestial sphere, extending nine degrees on each side of the ecliptic (see below). The name means "zone of animals." All the constellations in it save one, Libra, are figures of living creatures.

p. 106. *Azimuth.* The angular bearing of a heavenly body measured from the true meridian.

p. 106. *obnoxious.* Detrimental.

p. 107. *the Ecliptick.* The trace of the plane of the earth's orbit upon the celestial sphere.

p. 108. *Hart-Hall in Oxford.* Now incorporated with Magdalen Hall in Hertford College.

p. 108. *a Micrometer.* A micrometer eye-piece. An eye-piece such that any distance moved by the image of a star can be measured. This is a very early use of the name and device.

p. 108. *Managery.* Method of use.

p. 108. *Alcor.* The small companion star to *Mizar* in the tail of the Great Bear.

p. 108. *the Wayn.* Charles's Wain, or the Great Bear.

p. 108. *Vidit Alcor et non Lunam Plenam.* He sees *Alcor* but not the full moon.

p. 108. *Hevelius. John Hevelke* or *Hevelius* (1611–87), a German astronomer.

p. 109. *nice.* Sensitive, accurate.

p. 109. *the two Solstices.* The curve called the ecliptic cuts the celestial equator in two opposite points called the equinoxes. The two points on that equator half-way between the equinoxes are called the solstices.

p. 110. **Robert Hooke: "A Method for making a History of the Weather"**

p. 110. *Robert Hooke* (1635–1703). One of the ablest of the early Fellows of the Royal Society. Highly esteemed by Samuel Pepys and often mentioned in his diary.

p. 110. *that Dimension.* That level or position.

p. 110. *unwreathed.* Uncoiled.

p. 111. *the Cod.* The Pod.

p. 111. *the Action.* That Effect.

p. 112. *Ignes fatui.* Will o' the wisps.

p. 112. *shining exhalations.* Perhaps shooting stars and meteors.

p. 112. *always conversant.* Always to be found.

p. 114. *Reaches or Racks.* Streams of clouds in motion.

p. 114. *Supellex.* Equipment, apparatus for experimenting.

p. 114. **Martin Lister: "The Reason of the Ascent of the Quicksilver"**

p. 115. *Fretting, upon the Fret.* Vibration, continually vibrating.

p. 115. *purging of it.* Getting rid of the air.

p. 115. *the concave Figure.* Concave shape as viewed from inside the mercury. Nowadays one uses the phrase *convex meniscus* of the mercury, from the point of view of an external observer.

p. 115. *Superficies.* the plural 'surfaces.'

p. 115. *stagnates.* Stands, rests.

p. 116. *at a Stand.* Puzzled. The same expression is used about Christian in *Pilgrim's Progress.*

p. 116. *Zwelfer.* Probably Johann Zwelfer (1618–68), a famous pharmacist and physician in Vienna.

p. 116. *in Balneo.* Surrounded by a cold bath.

p. 116. *Lympha of the Blood.* Now called lymph. An alkaline, colourless liquid present in various tissues and organs of the body.

p. 116. *inspissated.* Caused to thicken.

p. 116. *Bornichius.* Probably Ole or Olaus Borch (1626–90), Professor of Chemistry at Copenhagen University.

p. 117. *we most converse with.* We have the most dealings with.

p. 117. *Sanguineous.* Perhaps warm-blooded, or simply "with blood in their veins."

p. 117. *the Glass which I whelmed upon them.* The glass which I placed over them.

p. 118. Robert Boyle: "New Experiments Physico-Mechanical Touching the Spring of the Air"

p. 118. *Robert Boyle* (1627–91). The Honourable Robert Boyle, the father of modern chemistry, was a son of Richard Boyle, the first and "great" Earl of Cork.

EXPERIMENT I.

p. 118. *explicable.* Note Boyle's use of explication for explanation.

p. 118. *your Lordship.* Boyle's own nephew, Charles Boyle, Viscount Dungarvan, eldest son of the second Earl of Cork.

p. 118. *parcel of air.* Portion of air.

p. 118. *resembled to.* Compared with, likened unto.

p. 118. *still endeavouring.* Continually endeavouring.

p. 119. *an endeavour outwards.* A force acting towards the outside.

p. 119. *loosely complicated.* Loosely intertwined.

p. 119. *to explicate.* To explain.

p. 120. *brandishing motion.* Oscillatory motion.

p. 120. *Annex.* Add.

p. 121. *wind-guns.* Air guns.

p. 121. *pneumatical engines.* Air pumps.

p. 121. *And did not their gravity hinder them . . . doth reach.* This rather obscure sentence means "If their gravity (weight) did not hinder them, there is no reason why the currents . . . should not rise much higher than the height of the atmosphere as deduced from the refractions of the sun and other stars, even if we take its highest estimated value."

p. 122. *a dry lamb's bladder.* One of Boyle's favourite vessels for holding air. He also uses a 'limber,' *i.e.*, flexible lamb's bladder.

p. 122. *recess.* Escape, withdrawal of air.

p. 122. *wave.* Waive, omit.

p. 122. *so nice an experiment.* So delicate an experiment.

p. 122. *the German experiment.* The Magdeburg experiment on atmospheric pressure, by Otto von Guericke.

p. 122. *the triers.* The experimenters. *Cf.* the Triers, the Ecclesiastical Commissioners set up by Oliver Cromwell for a very different purpose.

Experiment XXVII

p. 123. *Kircher.* Athanasius Kircher (1601–80), a Jesuit inventor of a magic lantern, mathematician, and writer on acoustics.
p. 123. *luciferous.* Luminous.
p. 124. *the second minutes.* The seconds.
p. 125. *reiterated.* Repeated.
p. 125. *close.* Dense.
p. 126. *the Torricellian experiment.* Evangelista Torricelli (1608–47), a brilliant pupil of Galileo. He invented a barometer by inverting a glass tube full of mercury, with the lower end in a pool of mercury.
p. 126. *a fitter.* A more suitable size.
p. 127. *the moveable stopple.* The removable stopper or cork.
p. 127. *junctures.* The joints, connexions.
p. 127. *These thoughts, my Lord . . .* This peculiar sentence may perhaps be expressed in the more modern form: "As I was saying, my Lord, we thought of trying the experiment of introducing compressed air into the receiver containing the source of sound, to find whether the sound heard was louder than in ordinary air. Since, however, we had no receivers in reserve, we dare not carry out the experiment for fear we might break our one and only receiver before we found how to compress the air in it."

p. 128. **Narcissus Marsh: "The Doctrine of Sounds"**
p. 128. *Narcissus Marsh.* Narcissus Marsh (1638–1713), born in Wiltshire. M.A. and D.D. (Oxon), Provost of Trinity College, Dublin (1678). Bishop of Ferns and Leighlin (1678), Archbishop of Cashel (1691), of Dublin (1694), and finally of Armagh (1703). A founder of the Dublin Philosophical Society. Wrote at least two papers on Acoustics. Assisted in translating the Old Testament into Irish.

p. 128. The peculiar numbering of the paragraphs is that of the original paper.
p. 128. *answerable whereunto.* Corresponding to which.
p. 128. *Ex parte Objecti.* At the source of light (literally, "from the side of the object").

p. 128. *Ex parte Organi vel Medii.* At the receiving end or in the medium between (literally, "from the side of the organ of seeing or the medium").

p. 128. *the begetting of Sounds.* The production of sounds.

p. 129. *Hollowing.* Shouting—*e.g.*, the Hallo of the huntsman.

p. 129. *Luring.* Making sounds to entice animals or birds.

p. 129. *the Systrum of the Egyptians.* A kind of a rattle used especially in the worship of the Egyptian deity Isis. Usually called *Sistrum.*

p. 129. *will be mightily conserved.* Will carry far without falling off in loudness.

p. 130. *Whispering places.* Whispering galleries. There is a fine example in the dome of St Paul's Cathedral.

p. 130. *rowls.* Rolls.

p. 130. *superficies.* Surface.

p. 131. *Otacousticks.* Ear trumpets. Devices of this kind were invented by Fellows of the Royal Society (Hooke) before 1667. Sprat (1667) calls them otacousticons.

p. 131. *the Visive Rays.* The rays by which an object is seen. The visual rays.

p. 132. *Purblind.* Partly blind.

p. 132. *Perspective glasses.* Hand telescopes, using one convex objective and one concave eyelens, as in the Galilean telescope, and giving an erect image. The Shepherds of the Delectable Mountains, in *Pilgrim's Progress* (1678), allowed Christian and Hopeful to look through such a perspective glass at the distant Celestial City.

p. 132. *Polyscopes.* This looks like a forestalling of the kaleidoscope, usually said to have been invented by Sir David Brewster in 1816. Pretty multiple patterns of a single object are produced by two inclined plane mirrors.

p. 132. *Acoustric.* The modern word is acoustic. Perhaps the 'r' is a misprint.

p. 132. *Stentoro-phonicon.* A speaking-trumpet or megaphone invented by Sir Samuel Morland. Satirical references to it are found in Shadwell's play *The Virtuoso*, and Mrs Aphra Behn's play *The Emperor of the Moon.*

p. 133. *Microphones.* This must be one of the earliest uses, if not the very first use, of this word, now so popular in the days of radio. Moderns, however, use the microphone at the transmitting end, or source of sound.

p. 134. *reflex'd.* Reflected.

p. 134. *the Corpus Obstans.* The body which reflects the sound. The body interposed in the path of the sound.

p. 136. *Looking-Glasses obverted.* Probably two parallel plane mirrors facing each other.

p. 136. *Problem I.* The modern invention of amplifiers solves this problem.

p. 136. *Problem II.* The telephone, transmitting either along wires or by wireless, solves this problem.

p. 137. *Problem III.* Problem III can now be regarded as solved by telephony through wires or by wireless telephony, unless intermediate observers listen in by the aid of special instruments. Problems I, II, and III are to be regarded as suggestions of topics for research, at the time they were propounded, not as announcements of results achieved.

p. 137. *A semiplane.* A horizontal cross-section of a set of waves of sound, leaving a point source. A diagram of this is shown on a neighbouring page of the *Philosophical Transactions.*

p. 137. *the Draught.* The Diagram.

p. 138. **Robert Hooke: " Of the Colours Observable in Muscovy Glass and other Thin Bodies "**

p. 138. *Muscovy Glass.* The mineral mica which occurs naturally in thin transparent sheets.

p. 138. *talk . . . spar and Kanck.* Apparently calc-spar or calcite. The term talc is now applied to a very different mineral, *viz.* a soft whitish substance consisting mainly of magnesium silicate. It is well known as tailor's chalk, French chalk, or steatite.

p. 138. *Selenitis.* Probably *gypsum.*

p. 139. *consecution.* Sequence, order of appearance.

p. 141. **Edmund Halley: " The Expansion of Several Fluids, in order to ascertain the Divisions of the Thermometer "**

p. 141. *Edmund Halley.* An eminent astronomer and physicist. Predicted the regular return of the comet called Halley's comet. See p. 208.

p. 141. *a large Bolt-head.* Usually a retort but apparently here a flask with a long, narrow neck.

p. 141. *the Augment.* The increase, the expansion.

p. 141. *a Skillet.* A pan, dish.

p. 141. *its Temper.* Its temperature.

p. 142. *apace.* Quickly.

p. 142. *attainted.* Attained.

p. 143. *which wanted.* Which fell short.

p. 144. *determined so nicely.* Determined so accurately.

p. 144. *rectify'd or dephlegmed.* Purified or concentrated by boiling off water.

p. 145. *that late honourable Patron of experimental Philosophy, Mr Boyle.* Robert Boyle died in 1691. This law was dated 1659.

p. 146. *the Scales of Heat and Cold.* An early use of the word 'scale' as applied to temperature.

p. 146. *M. Mariotte.* Edmé Mariotte (*c.* 1620–84), a French scientist who studied the physical laws of gases. What we call Boyle's law is called Mariotte's law in France. It came out in 1676.

p. 147. Robert Boyle: " A Dialogue on the Nature of Combustion "

p. 147. *Carneades.* The name of one of the interlocutors or supposed speakers taking part in the discussion. He usually addresses his friend Eleutherius.

p. 147. *I take then of copper filings . . .* The experiment of heating a mixture of copper filings, the so-called sublimate $HgCl_2$, and sal-ammoniac, NH_4Cl, probably gives metallic mercury by the reducing action of the copper on the mercuric chloride. The copper replaces the mercury in combination with chlorine. The presence of sal-ammoniac appears to be unnecessary.

p. 147. *a competent fire.* A fire giving sufficient heat.

p. 148. *true sulphur of Venus.* A kind of sulphur supposed to reside in the gem emerald.

p. 148. *occursions.* Acts of violent agitation.

p. 148. *a mountebank.* A comedian, comic entertainer.

p. 149. *the three principles.* Salt, sulphur, and mercury, the ingredients of bodies, according to an old view combated by Boyle.

p. 149. *aqua fortis* or *aqua regis.* Nitric acid or a mixture of three to four parts of hydrochloric acid and one of nitric acid.

p. 149. *colliquated.* Intimately fused together.

p. 149. *Helmont.* Probably Jean Baptiste van Helmont (1577–1644), a native of Brussels. A physician and one of the founders of modern chemistry. Invented the word 'gas.' Discovered some of the properties of carbon dioxide. Grasped the law of conservation of matter in particular cases.

p. 149. *alkahest.* The universal solvent.

p. 150. *effluviums.* Electrical fields as they are now called.

p. 150. *electrical concretes.* Substances capable of being electrified.

p. 150. Robert Boyle: " New Experiments Physico-Mechanical Touching the Spring of the Air "

p. 150. *Robert Boyle.* See p. 203.

EXPERIMENT XXXV

p. 150. *inquisitive.* Not busybodies, but simply of an inquiring mind.

p. 150. *a pipe of glass.* A glass tube. Sometimes Boyle prefers the expression, "a glass cane."

p. 150. *the plain of the water.* The level of the water surface.

p. 151. *as erected a posture.* As vertical a position.

p. 151. *exsuction.* Removal, sucking out.

p. 152. *mind.* Remind, inform.

p. 152. *useth to be concave.* Is usually concave.

p. 153. **Robert Boyle: " On Fluids "**

p. 153. 'The History of Fluidity' means the theory of fluids or the fluid state.

p. 153. *byass.* Bias.

p. 154. *superficies.* Surface.

p. 154. **Edmund Halley: The " Effects of Gravity in the Descent of Heavy Bodies and the Motion of Projects "**

p. 154. *Edmund Halley* (1656–1742), mathematician and astronomer. Friend of Isaac Newton. Visited St. Helena in 1677 and 1700.

p. 154. *Projects.* Projectiles.

p. 154. *Des Cartes.* René Descartes (1596–1650), a famous French philosopher and mathematician. The notion in the text is taken from his work *Principia Philosophiæ*, first published at Amsterdam in 1644.

p. 154. *in Libero Æthere.* In free space.

p. 155. *Dr Vossius.* The Latinized form of Isaac Vos (1618–89), scholar and canon of Windsor. Born at Leyden, Holland. Compiled a chronology of Scripture.

p. 155. *Principle.* Cause.

p. 155. *the Modus.* The way, the manner.

p. 156. *the Virtue of the Loadstone.* The force exerted by the loadstone.

p. 156. *to explain Ignotum per æque Ignotum.* To explain the unknown by the equally unknown.

p. 156. *Galilæus, Torricellius, Huyenius.* Latinized forms of the names of famous scientists known to us as Galileo (1564–1642), Torricelli (1608–47), and Huyghens (1629–95).

p. 156. *Mr Is. Newton.* Isaac Newton (1642–1727) was knighted by Queene Anne in 1705.

p. 158. **Edmund Halley: "A Theory of the Magnetical Variation"**

p. 158. *the Magnetical Variation.* Now called the angle of magnetic declination. The angle between the axis of a horizontal suspended magnet and a true north–south horizontal line.

p. 158. *this Table.* A table of values of the declination at various places and times is printed in the original paper.

p. 158. *the Tercera Islands.* Terceira is one of the islands of the Azores, in the Atlantic, west of Portugal, in longitude 27° 0′ West, and latitude 36° 42′ North.

p. 158. *Cape Frio.* In Brazil, seventy or eighty miles east of Rio de Janeiro. In longitude 42° 0′ West, and latitude 22° 55′ South.

p. 159. *Cape Comorin.* The extreme southern point of India. In longitude 77° 33′ East, and latitude 8° 4′ North.

p. 159. *Van Diemen's Land.* Tasmania, discovered by the Dutch sailor, Abel Tasman, in 1642, and called after the governor of the Dutch East Indies. Australia was also called by this name.

p. 159. *the Island Roterdam.* In the southern Indian Ocean, in longitude 84° 0′ East, latitude 20° 15′ South, according to Halley's own table. No island is to be found in that position in any atlas known to me. The island of Rodriguez is in longitude 64° East, latitude 20° South, so the 84° may be a misprint.

p. 159. *The Hon. Sir John Narborough.* According to Macaulay (*History of England*, Chapter III), there was a line of sea-dogs, of the right bulldog breed, each the cabin boy of his predecessor. Admiral Sir Christopher Mings had a cabin boy who became Sir John Narborough, who likewise had a cabin boy who became Admiral Sir Cloudesly Shovel. *Admiral Sir John Narborough* (1640–88). Like Nelson, a Norfolk man. Fought the Spanish, the Dutch, and also the corsairs of Tripoli.

p. 159. *at Baldivia.* Now Valdivia, a town in Chile. In longitude 73° 10′ West, latitude 39° 45′ South.

p. 159. *New Zealand.* This plain reference *c.* 1692 shows that New Zealand was discovered long before the time of Captain Cook.

p. 159. *Hound's Island.* Probably the coral atoll Dog Island or Puka Puka in latitude 15° South, and longitude 138° West, in the Tua Motu Archipelago of the South Pacific Ocean. Halley had probably read the account of its discovery by Willem Cornelison Schouten, a Dutch navigator, in 1616:

> The third of April, being Easter day, we were under fifteen degrees, twelve minutes (south latitude) at which time we had no

variation of Compasse, for the Needle stood right North and South. . . . A low Island not very great. . . . We held our course West, towards the Islands of Solomon, and called that Island, Dogs Island. [925 leagues from Peru.]

p. 160. *the River of Plata.* The large South American river in Argentina. Plata is the Spanish for Silver.
p. 160. *the Theory of Bond.* This is explained in the next few lines of the text.
p. 160. *the Pole Arctick.* The north geographical pole.
p. 161. *largest Dominions.* Longest range.
p. 161. *Hollandia nova.* Australia, particularly the northern parts. Ceylon was also called by this name in 1602.
p. 161. *the Magnetical Virtue.* The magnetic field.
p. 161. *which will still be greater.* Which will continually grow larger.
p. 162. *Japan, Yedzo.* The large northern island of Japan is called Yezo.
p. 162. *the Ethiopick Sea.* The South Atlantic.
p. 163. *Zelandia nova.* New Zealand.
p. 163. *the Access towards.* The approach to.
p. 165. *translated.* Moved from one place to another. A bishop is translated, say, from the see of St Asaph to that of Ely.
p. 165. *per saltum.* By a sudden event, by a jump.

p. 166. **William Petty: "An Apparatus to the History of the Common Practices of Dying"**
p. 166. *Sir William Petty, M.D.* (1623–87), ancestor of the Marquess of Lansdowne. A political economist, musician and physician.
p. 166. *The Bolonian Slate (called by some the Magnet of Light).* A grey or greyish-yellow variety of the mineral *barytes*, found near Bologna, Italy, and now called Bolognese Stone. It emits light when heated.
p. 166. *lucid.* Luminous.
p. 166. *superficies.* Surface.
p. 166. *suscipient.* Susceptible.
p. 167. *Humane.* Human.
p. 167. *Tiffanies.* The name tiffany has been applied to two sorts of cloth: (1) a kind of thin transparent silk, and (2) a transparent gauze muslin.
p. 168. *in a several way.* In a different or separate way.
p. 168. *Sal-Armoniac.* Sal-ammoniac.

p. 168. *Calamy.* Calamine, zinc carbonate, found in the form of greyish-white hexagonal crystals. The name was also applied to zinc silicate.

p. 168. *Hungarian vitriol.* A common kind of vitriol. Not sulphuric acid, but ferrous sulphate.

p. 168. *Cullen earth.* Cologne earth. A brown pigment obtained from lignite, a mineral originally found near Cologne.

p. 168. *Calces of Lead.* Oxides of lead; the singular is calx.

p. 168. *Ceruse and Minium.* White lead and red lead. Cinnabar, the red sulphide of mercury, was once called minium.

p. 168. *tinging.* In the sense of tinting, colouring.

p. 168. *variously.* In various ways.

p. 168. *Verdeter.* A kind of paint of a green, bluish-green or blue colour, made by adding whiting or chalk to copper nitrate.

p. 168. *the Lakes.* A lake was originally a pigment of a reddish made from lac, a dark-red Oriental resin. Later the sense was extended to a class of pigment made by mixing various colouring matters with metallic oxides.

p. 168. *Fæculæ.* Sediments at the bottom of vats.

p. 168. *Gambrugium.* Gamboge.

p. 168. *Indico.* Indigo.

p. 168. *clammy.* Viscous.

p. 168. *Oyl of Spike.* Oil of Spike is still the pharmaceutical name for oil of lavender.

p. 168. *Watering of Tabbies.* Watering in the textile sense, as in the phrase 'watered silk.' The name comes from Attabiy, a suburb of Baghdad where they were once manufactured.

p. 168. *Lixiviums.* Lyes. Solutions of alkalis obtained from wood ash, in water.

p. 168. *Macerations.* The process of extracting juices of plants by steeping in a liquid is called maceration.

p. 168. *Copperas.* This name was formerly applied to at least three separate things: the crystals of (1) blue copper sulphate, (2) green ferrous sulphate, and (3) white zinc sulphate. Here perhaps the second is meant.

p. 169. *Bow dye.* A scarlet dye, called after Bow, or Stratford-le-Bow, in East London. Chaucer refers to Stratford-atte-Bowe.

p. 169. *A fit Menstruum.* A suitable solvent.

p. 169. *to scour the Sordes, which may interpose between the Coloranda and the Dying Stuff.* To remove the dirt or filth, which may separate the material to be dyed, from the dye.

p. 170. *Fulling-mills.* Dye works.

p. 170. *intenerate.* Soften.
p. 170. *Grograins.* Grograms. A cloth made of silk, mixed with wool and mohair.
p. 170. *the Blue fat.* The blue vat.
p. 170. *Incarnadives.* Red dyes. Usually spelt 'incarnadines.'
p. 170. *a Vinculum.* A bond, link.
p. 170. *Aculei.* O.E.D. defines Aculei as prickles or stings. Perhaps here it means small crystals, which are rough or pointed.
p. 171. *levigates the Superficies.* Whitens the surface.
p. 171. *Aqua fortis.* Nitric acid.
p. 172. *Argol.* The white substance deposited by fermented wines on the sides of casks. It is a crude form of potassium bitartrate, and when purified gives cream of tartar.

p. 172. **Thomas Henshaw: "The Manner of Making Salt-Peter"**

This memoir had the honour of being attacked by Henry Stubbe in *Animadversions upon the History of making Salt-Peter which was Penned by Mr Henshaw* (1669). Stubbe was very severe upon Henshaw. Among other remarks he said: "The History of Salt-Peter hath so many defects in it, that I wonder any one should offer such an account to them"; and again: "The Narration is not only imperfect, but in many parts false." There were two Henshaws, brothers, Thomas Henshaw (1618–1700) and Nathaniel Henshaw, M.D. (1628–73), both among the earliest Fellows of the Royal Society. The writer of the present article was Thomas Henshaw. He was a friend of John Evelyn.

p. 172. *On the outside downward.* Near the bottom on the outside.
p. 172. *Peter-earth.* Crude mineral nitre.
p. 172. *a rundle of Wicker.* A round sheet or disc of wicker work.
p. 172. *this Lixivium.* The filtrate, liquid which drains through.
p. 173. *for one of the Profession.* For a professional manufacturer of salt-petre.
p. 173. *Fewel.* Fuel.
p. 173. *the brasen Scummer.* The bronze ladle for skimming.
p. 174. *ready to shoot into Peter.* Ready to crystallize out into salt-petre.
p. 175. *common or cubic Salt.* Common salt crystallizes in the form of cubes.
p. 175. *to gravulate.* To form, to crystallize out.

p. 175. *Nitrum obsonia alba et deteriora reddit Olera viridiora.* Philemon Holland translates this passage from Pliny: "As for cates and meats, if they bee powdred withall, they will look white and be worse for it: whereas all woorts for pot or sallad, will seeme the greener." 'Woorts' or 'worts' are plants.

p. 176. **William, Viscount Brouncker: "Experiments of the Recoiling of Guns."**

p. 176. *The lord Brouncker.* William, Viscount Brouncker (*c.* 1620–1684), a mathematician and first President of the Royal Society. Mentioned in the diaries of Pepys and Evelyn.

p. 176. *this College.* Gresham College, in Bishopsgate, London, was the early home of the Royal Society.

p. 176. *some literal impulse.* That is to say, some definite, appreciable impulse.

p. 176. *sensible.* Appreciable, perceptible.

p. 177. *his Majesty and his Highness Royal.* Charles II, and probably his brother, James, Duke of York, who patronized the work of the virtuosi.

p. 177. *Engine.* Machine, apparatus; in this case a small cannon.

p. 177. *the Tilt-yard Gallery.* Adjoining Whitehall Palace.

p. 178. **Prince Rupert of Bavaria: "To Make Small Shot of Different Sizes."**

p. 178. *Prince Rupert* (1619–82), noted for headlong cavalry charges in the Civil War. After the Restoration he dabbled in science, and is credited with some inventions.

p. 178. *at top.* On the top.

p. 178. *Auripigmentum.* Orpiment, the yellow sulphide of arsenic.

p. 179. *pour on new Lead still.* Keep on pouring lead.

p. 179. **Robert Hooke, "A Prediction of Artificial Silk."**

p. 179. *Robert Hooke.* See p. 202.

p. 179. *the Silk-worm wire-draws his clew.* "The silkworm draws out its fibre."

p. 179. *I have my aim.* "I shall achieve my aim, I shall be satisfied."

p. 180. **Thomas Salusbury: "Aristotle and the Telescope"**
From *The Systeme of the World in Four Dialogues* (1661). This was

a translation from the Italian of *Dialogi Di Galileo Galilei Linceo . . . dove ne i Congressi di quattro Giornate si discorre sopra i Due Massimi Sistemi Del Mondo Tolemaico, E Copernicano* (1632). The literal translation of the Italian title is "Dialogues by Galileo Galilei, Fellow of the Academy of the Lynxes, in which, in a four-day conference, the two greatest systems of the universe, Ptolemaic and Copernican, are discussed." This is the work which in 1632 brought Galileo into trouble with the Inquisition and the book was placed on the Index Liberum Prohibitorum, the black list of the Roman Catholic Church.

Thomas Salusbury published several translated books between 1660 and 1665. In this extract from Dialogue II the interlocutors or speakers are Salviati, Sagredo, and Simplicio. Salviati takes the Copernican point of view, and has been supposed to represent Galileo himself. Simplicio takes the old-fashioned Aristotelian point of view, and Sagredo, a neutral, acts as chairman.

p. 180. *diaphanous.* Transparent.

p. 180. *Statua's.* Statues.

p. 180. *the prophesies of Abbot Joachim.* Joachim de Fiore (1132–1202), an Italian Cistercian monk. Abbot of Corazzo (1178–88). Head of a 'mystical' school of interpretation of Scripture. A prophecy about the Popes, published under his name, had great success.

p. 180. *the Heathen Oracles.* The oracles at Delphi and elsewhere used to give ambiguous answers, and the double meaning often deceived the recipients.

p. 180. *the predictions of the Genethliacks.* Ancient astrologers, who pretended to tell the fortunes of a person, from the position of the stars at the moment of his birth. From γευέθλη, a Greek word meaning 'birth.'

p. 182. **Philemon Holland: "The Sundry Kinds of the Loadstone, and the Medicines thereto appending," from "The Historie of the World, commonly called The Naturall History of C. Plinius Secundus."**

p. 182. *Philemon Holland* (1552–1637). The "translator generall in his age." Faithfully translated works by Livy, Marcellinus, Pliny, Plutarch, Suetonius, Xenophon, Camden, and others.

p. 182. *trauelled.* Travailed, gone to much trouble.

p. 182. *the load-stone Sideritis, others Heracleos.* Probably the modern magnetite, a ferromagnetic, dark-brown, heavy mineral with the composition Fe_3O_4.

p. 182. *Nicander.* A Greek poet and physician, born about 150 B.C. Wrote on agriculture, remedies for wounds, and poisons.

p. 182. *the mountaine Ida.* Probably a range of mountains in the old province of Mysia, in Asia Minor. The loadstone is not associated with the more famous Mount Ida, in Crete.

p. 182. *the yron picke or graine of his staffe.* The iron tip or point of his shepherd's staff.

p. 182. *Sotacus.* A Greek philosopher flourishing in the fourth century B.C. An able mineralogist. Quoted by Pliny seven times.

p. 183. *Æthyopia.* A country in Africa, south of Egypt, rather vaguely including Abyssinia, Nubia, and the Sudan. Popularly regarded as the native land of negroes. In this translation of Pliny, Holland spells the word with an initial diphthong the first time, an initial E the second and third times, and Ehyopia the fourth time.

p. 183. *lake Bœbeis.* A lake in Macedonia, now Thessaly, north of Greece, at the foot of Mount Pelion. Also mentioned by Ovid. Now called lake Carla.

p. 183. *Echium, a town of Bœotia.* Bœotia was city state in ancient Greece, north-west of Athens. The country of Pindar and Epaminondas. The capital was Thebes.

p. 183. *Alexandria, in the region of Troas.* A town, often called Alexandria Troas or simply Troas, in north-west Asia Minor. The Troas of the Acts of the Apostles (16. 8 and 20. 5).

p. 183. *Natolia.* A province of Asia Minor.

p. 183. *simply the best.* By far the best.

p. 183. *Zimiri.* A district forming part of Ethiopia.

p. 183. *the sanguine load-stone, called Hæmatites.* Sanguine means blood red and hæmatites means blood stone.

p. 183. *the stone Theamedes that will abide no yron.* I have not yet been able to identify this mineral.

From **"The first inuention of glasse, and the manner of making it."**

p. 184. *Iurie.* Jewry, Palestine.

p. 184. *a meere named Cendeuia; out of which the river Belus is thought to spring.* Now a marsh called El Herdane, at the foot of Mount Carmel, near the coast of Palestine. The river Belos, now called Nahr Namen, still flows from it.

p. 184. *Ptolemais.* A town on the coast of ancient Palestine. On the site of the modern Acre.

p. 184. *mordacity.* The biting taste.
p. 184. *to seeth.* To boil, stew.
p. 184. *the very sands.* The sands themselves.
p. 184. *treuets.* Tripods to hold the stew pans.
p. 185. *the Load-stone . . . is thought naturally to draw the liquor of glasse vnto it, as well as yron.* This would appear to have been a mistaken observation. The most delicate modern instruments are required to detect the minute force exerted by a magnet on molten glass. If a ferromagnetic such as iron filings or dust were mixed with the glass, some such attraction as that mentioned by Pliny would occur.
p. 185. *Ophyr.* The same Ophir as the one mentioned in the Old Testament, from whence Solomon obtained supplies of gold. See I Kings ix, 28, xxii, 48, Job xxii, 24, xxviii, 16, and two other places. Its site is not quite certain, but modern scholars think it was in Africa, between the upper Nile and the Red Sea coast of Abyssinia.
p. 185. *parcells.* Pieces of glass.
p. 185. *the riuer Vulturnus.* The principal river in the Italian province of Campania. Now called the Volturno. About 40 km. north-west of Naples. Flows to the south-west.
p. 186. *between Cumes and the lake Lucrinus.* Cumae was a town, the most ancient Greek colony in Italy, in the Italian province of Campania, 15 to 20 km. west of Naples, half a mile south of lake Avernus and near cape Misenum. Lake Lucrinus is now called lake Lucrino or Maricello.
p. 186. *this bruit hath run currant.* A report to this effect has been going around.
p. 186. *Obsidiana.* Obsidian is a kind of volcanic glass, a semi-gem, of variable composition, generally containing either fused felspars or alkaline silicates. Resembles what is now called black glass.
p. 186. *the sight therein is but thick and duskish.* This is a charming paraphrase of the scriptural "Now we see through a glass darkly," of I Corinthians xiii, 12.
p. 187. *Iaiet.* Jet.
p. 187. *Cassidony.* Chalcedony, a precious stone, usually of a milk-white colour.
p. 188. *rundles.* Round discs or pieces, counters.

APPENDIX

THE CONTEMPORARY ATTITUDE TOWARDS SCIENCE

In the course of the seventeenth century there appeared a number of plays, poems, and prose works, satirizing false science, and satirizing, attacking, or eulogizing true science.

The famous play called *The Alchemist*, by Ben Jonson (*c.* 1573–1637), produced about 1610, is a comedy of manners satirizing not genuine science but the alchemy of quacks and tricksters such as the Dees and Kellys of the period. Written not long after the publication of Gilbert's *De Magnete* (1600), it is not surprising that it contains a reference to magnetism in Act I, Sc. 3:

> And,
> Beneath your threshold, bury me a load-stone,
> To draw in gallants, that wear spurres: the rest,
> They'll seeme to follow.

Some of the technical jargon used in connexion with the transmutation of metals has a legitimate meaning in modern chemistry. Thus in Act II, Sc. 5 we have a speech by Subtle, the pseudo-scientist:

> . . . Stand you forth, and speake to him
> Like a philosopher: answer i' the language.
> Name the vexations, and the martyrizations
> Of mettalls, in the worke.

The assistant, Face, replies

> . . . Sir, putrefaction
> Solution, ablution, sublimation,
> Cohobation, calcination, ceration, and
> Fixation.

Very few of these words are unintelligible to moderns. We are told in the text that cohobation is "the powring on your aqua regis, and then drawing him off." In the same act and scene there is a section with a very up-to-date flavour.

> SUBTLE: . . . And, what's your mercury?
> FACE: A very fugitive, he will be gone, sir.
> SUBTLE: How know you him?
> FACE: By his viscositie,
> His oleositie and his suscitabilitie.

Remarkable words for 1610!

Act IV, Sc. 3, has a topical reference to modern methods:

> SUBTLE: I'll ha' you to my chamber of demonstrations,
> Where I'll show you both the grammar and logick,
> And rhetorick of quarrelling; my whole method,
> Drawne out in tables: and my instrument,
> That hath the severall scale upon 't. . . .

In 1676 Thomas Shadwell (*c.* 1642–92), Dryden's rival, brought out a play called *The Virtuoso*, a vigorous satire of the Royal Society and its labours, but particularly of Sir Robert Howard (1626–98), one of the Fellows, and at the same time a mediocre dramatist himself. Howard is said to have been a pretender to all manner of arts and sciences.

In the comedy his counterpart is the virtuoso, Sir Nicholas Gimcrack, who practises swimming on dry land. Another time he counter-transfuses blood between a spaniel and a bulldog, thus causing the animals to exchange characteristics. He studies the thirty-six species of English spiders and has kept a tame one called Nick, which answers to his name, and follows his master about the house. Agents in the country send him bottles of fresh air from Newmarket, Norwich, and other places, including a special brand from Bury St Edmunds. These samples he releases in his house to refresh his friends.

He studies the phosphorescent light from putrefying objects and has read a Geneva Bible by the gleams from a decaying leg of pork. He is about to illuminate his study with no other lamps but glow-worms. He has made an effective ear-trumpet, which he calls a Stentrophonical Tube, for the better study of acoustics. Through his telescope he has closely observed the moon, seeing on it valleys, seas, lakes, elephants, and camels, and especially an ambitious prince who aims at universal monarchy.

A rabble of ribbon weavers attacks his house on a report that he has invented a power loom, or engine-loom as they call it. In vain he tells them that "we Virtuosos never find out anything of use, 'tis not our way," and he is about to be hanged when rescued by friends. All these effective hits were based on the regular employments of the first Fellows of the Royal Society, as reported by Sprat or the early *Philosophical Transactions*.

In this play too we note the phrase "Tace is Latine for a candle," so frequently quoted by Scott in the "Waverley Novels." It is put in the mouth of Sir Samuel Hearty, a boisterous character, who has little knowledge of the meaning of words. Shadwell also satirized Sir Robert Howard as Sir Positive At-All in *The Sullen Lovers* (1668), and some of the shafts in *The Virtuoso* are directed at John Dryden as a dramatist, especially as the author of *Aureng-Zebe*.

When the great Italian Galileo Galilei (1564–1642) turned the newly invented refracting telescope towards the skies in 1610, making a number of first-rate discoveries about the planets and stars, he made a special study of the features of the moon's surface. His researches made a profound impression on the thinking portion of the people of Europe. John Milton (1608–74) visited Galileo in the year 1638, and looked through the "optic-glass" at the heavens. Hence in *Paradise Lost*, Book I:

> . . . Like the moon, whose orb
> Through optic glass the Tuscan artist views
> At ev'ning, from the top of Fesole
> Or in Valdarno, to descry new lands,
> Rivers or mountains in her spotty globe.

Paradise Lost is permeated with allusions to the Ptolemaic astronomy, as if Milton had not accepted the views of Copernicus.

In 1638 there was published a work by Francis Godwin (1562–1633), Bishop of Llandaff and later of Hereford, called *The Man in the Moone, or a Discourse of a Voyage thither by Domingo Gonsales, the Speedy Messenger*. This is a piece of brilliant non-satirical prose fiction, anticipating H. G. Wells or Baron Munchausen. Gonsales is represented as being drawn up to the moon by twenty-five geese, in a kind of aerial car or parachute, and we are told that "the covey carried him along lustily." A second edition came out in 1657. It is supposed that John Wilkins (1614–72), Bishop of Chester, and a founder of the Royal Society, obtained from the Discourse several hints for his first publication, a serious prose essay called *The Discovery of a World in the Moone, or a Discourse tending to prove that 'tis probable there may be another Habitable World in that Planet* (1638). To the third edition of 1640 he added *A Discourse concerning the Possibility of a Passage thither.* Wilkins's work was honoured by a French translation as *Le Monde dans La Lune* (1665) by Le Sieur de la Montagne. The French Savinien de Cyrano Bergerac (1620–55), better known from the modern Rostand play as Cyrano de Bergerac, the man with the monstrous nose, is believed to have borrowed ideas from Wilkins which he used in *L'Autre Monde ou les Etats et Empires de la Lune* (1649). This was "done into English" in 1659 as *Selenarchia, or the Government of the World in the Moon*, by Thomas St Serf, and again "newly Englisshed" by A. Lovell in 1687 as *The Comical History of the States and Empires of the Worlds of the Moon and Sun.* Jonathan Swift (1667–1745) appears to have obtained material for the third book of *Gulliver's Travels* either directly from Wilkins or indirectly from him via de Bergerac.

In the early days of the Royal Society, and particularly after it received its charter in 1662, the virtuosi—*i.e.*, the Fellows among whom Wilkins and Sir Paul Neale (or Neile) (*c.* 1613–86) were prominent—devoted much attention to observations of the moon through improved telescopes. Neale seems to have made a fifty-foot refractor and with it perhaps thought he saw rivers and trees on the moon's surface. At any rate, a friendly hand, probably that of Joseph Glanvill (1636–80), one of the early Fellows, wrote *A Humorous Ballad of Gresham College*, praising "that choice Company of Witts and Philosophers who meet on Wednesdays weekly att Gresham Colledg." In this he gently rallied Wilkins, Neale, and the rest on their researches.

Other contemporaries did not regard these activities so favourably; Samuel Butler (1612–80), the author of *Hudibras*, wrote *The Elephant in the Moon*, a satirical poem, in two distinct versions with different metres. One version consisted of 520 lines in 'short' verse, the other 538 lines in 'long' verse, but they were not published until 1759. In each case the philosophers are shown to us in the act of watching through a telescope a fierce battle between the Privolvans and the Subvolvans, inhabitants of our satellite. Suddenly an elephant is seen in the fray, but after exciting much interest is, alas, discovered to be a mouse in the tube of the instrument.

Here is the beginning of the version in long verse:

A Virtuous, learn'd Society, of late
The Pride and Glory of a foreign State,
Made an Agreement on a Summer's Night,

To search the Moon at full, by her own Light;
To take a perfect Invent'ry of all
Her real Fortunes, or her Personal;
And make a geometrical Survey,
Of all her Lands, and how her Country lay,
As accurate, as that of Ireland, where
The sly Surveyor's said t'have sunk a Shire:
T'observe her Country's Climate, how 'twas planted,
And what she most abounded with, or wanted;
And draw Maps of her prop'rest Situations
For settling, and erecting new Plantations;
If ever the Society should incline
T' attempt so great, and glorious a Design:
A Task in vain, unless the German Kepler
Had found out a Discovery to people her,
And stock her Country with Inhabitants
Of military Men, and Elephants.
For th' Ancients only took her for a Piece
Of red-hot Iron, as big as Peloponese,
Till he appear'd; for which, some write, she sent
Upon his Tribe as strange a Punishment.
This was the only Purpose of their Meeting,
For which they chose a Time, and Place most fitting;
When, at the Full, her equal Shares of Light
And Influence were at their greatest Height.
And now the lofty Telescope, the Scale,
By which they venture Heav'n itself t'assail,
Was rais'd, and planted full against the Moon . . .

Butler also 'quizzed' the scientists by his description of A Virtuoso in his prose work *Characters*. In a third unfinished poem of 104 lines, called *A Satire on the Royal Society and Sir Paul Neal*, Butler has some very pertinent lines on the nature of magnetism, still an unsolved mystery:

Or what's the strange magnetic Cause
The Steel or Loadstone's drawn or draws,
The Star, the Needle, which the Stone
Has only been but touched upon?
Whether the North-Star's influence
With both doth hold Intelligence
(For red-hot Ir'n, held tow'rds the Pole,

Turns of itself to 't, when 'tis cool)
Or whether Male and Female screws
In th' Ir'n and Stone th' Effect produce? (ll. 53–62)

and he hints at a theory of crystal structure:

To explicate, by subtle Hints,
The Grain of Diamonds and Flints. (ll. 99–100)

The second part of *Hudibras* itself contains as a principal character the astrologer Sidrophel, who is the object of many of the shafts of wit therein. Authorities differ as to the person aimed at. Some think it to be a real astrologer of the period, William Lilly. Others suppose rather that it is Sir Paul Neale (or Neile), the eminent virtuoso, mentioned above.

The benefits conferred and likely to be conferred on humanity by scientific research are recognized by the great poet John Dryden (1631–1700) in the poetical epistle to his friend Dr Walter Charleton (1663). Dryden himself had been elected a Fellow of the Royal Society on November 26, 1662. This poem contains the lines:

Among the asserters of free reason's claim,
Our nation's not the least in worth or fame.
The world to Bacon does not only owe
Its present knowledge, but its future too.
Gilbert shall live, till loadstones cease to draw,
Or British fleets the boundless ocean awe,
And noble Boyle, not less in nature seen,
Than his great brother, read in states and men.
The circling streams, once thought but pools, of blood
(Whether life's fuel, or the body's food)
From dark oblivion Harvey's name shall save;
While Ent keeps all the honour that he gave.

On the other hand, Dryden produced in 1668 a comedy called *An Evening's Love* or *The Mock Astrologer*, founded on *Le feint Astrologue* of Thomas Corneille. The satire is very mild and is aimed at the pseudo-scientists. A witty pantomimic farce, with more bite in it than *The Mock Astrologer*, was Mrs Aphra Behn's *The Emperor of the Moon* (1687), based on the French *Arlequin Empereur dans La Lune*, of unknown authorship, but produced by D. Biancolelli, and published in 1684. This play contains hits at Gonsales and Wilkins but more especially at the Rosicrucian order of mystics. One scene is reminiscent of Butler, for in it Dr Balardo, while gazing through a twenty-foot telescope, is made to see an emperor on his throne, nymphs and clouds, "peopling the vast regions of the air." These are merely figures drawn on a sheet of glass, with a lamp behind it, which a mischievous person holds in front of the objective lens of the great telescope, in full view of the audience but unseen by the doctor. This makes a vigorous stroke at the optical researches of the Royal Society. From the pay-box point of view the play was a failure.

Harvey's scientific work was commemorated and described in the *Ode upon Dr Harvey* (1657) by the poet Abraham Cowley (1618–67). This ode is in the peculiar metre favoured by that poet. Two stanzas explain the principle of the circulation of the blood, a third relates to embryology, and two others summarize and comment upon the whole. Cowley also wrote another ode *To the Royal Society* (*c.* 1661), and this was prefixed to Spratt's history of that body (1667). In this, having likened Francis Bacon unto a Moses who had led mankind out of the barren wilderness of misdirected effort, he went on to compare the Royal Society to the little band whom Gideon led against the Midianites. Having broken their empty pitchers of old false methods and picked up the lamps of the new true ones, they were to sound trumpets of victory over the secrets of the universe. In 1661 Cowley also wrote in prose *A Proposition for the Advancement of Experimental Philosophy*. He himself was a founder member of the Royal Society.

Thomas Hobbes (1588–1679), the famous philosopher, had definite views on certain branches of knowledge such as the nature of matter, optics, and even psychology, but he did not seize on the new method and experiment for himself. His doctrines are explained in *The Elements of Law, Natural and Politique* (1640), *Tractatus Opticus* (1644), and *De Homine* (1658). He attacked Robert Boyle's views on gases in *Dialogus Physicus, sive de Natura Æris* (1661), and Boyle replied in *Examen of Mr Hobbes*. A further work of this kind, by Hobbes, was *Problemata Physica* (1662). Broadly speaking, Hobbes explained natural phenomena on geometrical and mechanical principles.

Sir Matthew Hale (1609–76), Lord Chief Justice of the King's Bench, had the old-fashioned views on natural phenomena. He published among many other works (1) *An Essay touching the Gravitation or Non-Gravitation of Fluid Bodies, and the Reasons thereof* (1673 and 1675), (2) *Difficiles Nugæ; or Observations touching the Torricellian Experiment, and the various Solutions of the same, especially touching the Weight and Elasticity of the Air* (1674), and (3) *Magnetismus Magnus;*

or Metaphysical and Divine Contemplations on the Magnet or Loadstone (1695). None of these possess any scientific value. The second was an arraignment of the views of Boyle and others on the weight and 'spring' of air, and led to a criticism by Dr Henry More and a reply by Hale.

As we have seen, the Royal Society was not allowed to boast of its methods and successes unopposed. It aroused the ire of the followers of Aristotle, especially at the universities, and, perhaps, of some of the Fellows of the Royal College of Physicians and apothecaries who feared their authority might be undermined. They tried to engage public favour by asserting that the new philosophy would destroy the Church of England and cause a return to Roman Catholicism, the universal religion proposed by the Italian writer Campanella. One of the first Fellows of the Royal Society, the Joseph Glanvill (1636–80) mentioned above, and Rector of the Abbey Church at Bath, published a prose work *Plus Ultra; or the Progress and Advancement of Knowledge since the Days of Aristotle* (1668) against the Aristotelian Robert Crosse, Vicar of Chew Magna, in Somerset. Henry Stubbe (1632–76), a physician at Warwick, a person whose head was "direct Carrot," according to his opponent, took up the cudgels against Glanvill and the Royal Society. It is possible that Stubbe was hired by Dr Baldwin Hamey of the Royal College of Physicians or by Dr Fell of Oxford (the Dr Fell of our nursery rhymes), but the evidence is not convincing. He disclaimed such charges, but his contemporaries suspected his disinterestedness. In any case he replied to Glanvill in *The Plus Ultra reduced to a Non Plus: or a Specimen of Some Animadversions upon the Plus Ultra of Mr Joseph Glanvill* (1670).

Stubbe was the abler man, skilled in all the weapons of controversy, but defending the weaker cause. Glanvill was conceited and frequently exposed a flank to the enemy, who took due advantage thereof. These two engaged in a war of pamphlets and letters. Stubbe also wrote a book *Legends no Histories* (1670) against Sprat's *History of the Royal Society* (1667). In the end, Stubbe was accidentally drowned in 1676 near Bath, and Glanvill preached his funeral sermon, thereby getting the last word. On the whole Stubbe caused no small embarrassment to the Royal Society. Another scholar, Meric de Casaubon (1599–1671), opposed Glanvill and Sprat in *Of Credulity and Incredulity* (1668) and a *Letter to P. du Moulin* (1669), claiming that the ancients had made many valuable inventions as well as the moderns.

Bearing on the scientific literature of the seventeenth century, though published later, we may briefly mention the *Poem Sacred to the Memory of Sir Isaac Newton* (1727), by James Thomson (1700–48). This describes Newton's achievements in splendid language. Book III of the immortal *Gulliver's Travels* (1726) is a biting prose satire on the early Royal Society, rendered even more spicy by the fact that one or two of the fanciful Laputan research projects, such as the solidification of air and the extraction of nitrogen from it, have since been realized experimentally.

MODERN WORKS RELATING TO BRITISH SEVENTEENTH-CENTURY SCIENCE

General Science

BROWN, E. W., and others: *The Development of the Sciences* (Yale, 1923).

BUTTERFIELD, H.: *The Origins of Modern Science, 1300–1800* (London, 1949).

BURTT, E. A.: *The Metaphysical Foundations of Modern Physical Science* (New York and London, 1925).

CLARK, G. N.: *Science and Social Welfare in the Age of Newton* (Oxford, 1937).

CLARK, G. N.: *The Seventeenth Century* (Oxford, 1947).

DAMPIER, W. C.: *A History of Science* (Cambridge, third edition 1942).

GREGORY, R.A.: *Discovery or The Spirit and Service of Science* (London, 1916).

GUNTHER, R. T.: *Early Science in Cambridge* (Oxford, 1937).

JEANS, SIR JAMES H.: *The Growth of Physical Science* (Cambridge, 1947).

LYONS, SIR HENRY: *The Royal Society, 1660–1940* (Cambridge, 1944).

MEES, C. E. K.: *The Path of Science* (London, 1946).

ORNSTEIN, M.: *The Role of Scientific Societies in the Seventeenth Century* (Chicago, 1938).

PLEDGE, H. T.: *Science since 1500, A Short History of Mathematics, Physics, Chemistry and Biology* (London, 1939).

SAXTON, G.: *The History of Science and the New Humanism* (Cambridge, Mass., 1937).

SCHUSTER, A. AND SHIPLEY, A. E.: *Britain's Heritage of Science* (London, second edition, 1920).

STIMSON, DOROTHY: *Scientists and Amateurs, A History of the Royal Society* (London, 1949).

SYFRET, R. H.: "The Origins of the Royal Society," *Notes and Records of the Royal Society*, Vol. V (1948), No. 2.

SYFRET, R. H.: "Some Early Critics of the Royal Society," *Notes and Records of the Royal Society*, Vol. VIII (1950), No. 1.

TAYLOR, F. S.: March of Mind. *A Short History of Science* (London, 1939).
WIGHTMAN, W. P. D.: *The Growth of Scientific Ideas* (London, 1950).
WOLF, A.: *A History of Science, Technology and Philosophy in the Sixteenth and Seventeenth Centuries* (London, 1938).

Biology, Botany, Physiology, and Zoology

BAYON, H. P.: "William Harvey, Physician and Biologist," *Annals of Science*, Vols. III and IV (London, 1938–40).
COMRIE, J. D.: *History of Scottish Medicine* (London, 1931). 2 vols.
FOSTER, SIR MICHAEL: *Lectures on the History of Physiology during the 16th, 17th, and 18th Centuries* (Cambridge, 1901).
GUNTHER, R. T.: *Early British Botanists and their Gardens* (Oxford, 1922).
GUTHRIE, D.: *A History of Medicine* (London, 1945).
MIALL, L. C.: *The Early Naturalists, Their Life and Work (1530–1789)* (London, 1912).
NORDENSKIÖLD, E.: *The History of Biology: A Survey*, translated by L. B. Eyre (London, 1928).
OLIVER, F. W. (editor): *Makers of British Botany* (Cambridge, 1912).
RAVEN, C. E.: *John Ray, Naturalist: His Life and Works* (Cambridge, 1942).
REED, H. S.: *A Short History of the Plant Sciences* (Waltham, Mass., 1942).
REYNOLDS, G. J.: *A History of Botany in the United Kingdom* (London, 1914).
SINGER, C. J.: *A Short History of Medicine* (Oxford, 1928).

Chemistry, Geology, and Physics

ADAMS, F. D.: *The Birth and Development of the Geological Sciences* (London, 1939).
ANDRADE, E. N. DA C.: *Isaac Newton* (London, 1950).
BELL, A. E.: *Christian Huygens and the Development of Science in the Seventeenth Century* (London, 1948).
BREWSTER, SIR D.: *Memoirs of the Life, Writings and Discoveries of Sir Isaac Newton* (Edinburgh, 1855).
BROAD, C. D.: *The Philosophy of Francis Bacon* (Cambridge, 1926).
BRODETSKY, S.: *Sir Isaac Newton* (London, 1927).
BROWN, G. B.: *Science: Its Method and Philosophy* (London, 1950).

CAJORI, F.: *A History of Physics in its Elementary Branches* (New York, 1929).
COOPER, L.: *Aristotle, Galileo and the Tower of Pisa* (Oxford, 1935).
DAVIS, F. L.: "Boyle's Conception of Elements Compared with That of Lavoisier," *Isis*, Vol. 16, 1931.
FERCHL, F. AND SÜSSENGUTH, A.: *Pictorial History of Chemistry* (London, 1939).
HART, I. B.: *Makers of Science* (Oxford, 1923).
HOLMYARD, E. J.: *Chemistry to the Time of Dalton* (Oxford, 1925).
LODGE, O. J.: *Pioneers of Science* (Kepler, Galileo, Descartes, Newton) (London, 1892).
LOWRY, T. M.: *Historical Introduction to Chemistry* (London, 1936).
MASSON, F.: *Robert Boyle* (London, 1914).
MCCOLLEY, G.: "The Seventeenth Century Doctrine of a Plurality of Worlds," *Annals of Science*, Vol. I, 1936.
MEYER, E. VON: *A History of Chemistry from Earliest Times to the Present Day*, translated by G. McGowan (London, 1891).
MORE, L.: *Life and Works of the Honourable Robert Boyle* (Oxford, 1945).
PARTINGTON, J. R.: *A Short History of Chemistry* (London, 1948).
SNOW, A. J.: *Matter and Gravity in Newton's Physical Philosophy* (Oxford, 1926).
STILLMAN, J. M.: *The Story of Early Chemistry* (New York, 1924).
STONER, G. B.: "The Atomistic View of Matter in the XVth, XVIth, and XVIIth Centuries," *Isis*, Vol. 10, 1928.
THORPE, SIR EDWARD: *History of Chemistry* (London, 1909).
ZITTEL, K. A. VON: *History of Geology and Palæontology*, translated by M. M. Ogilvie-Gordon (London, 1901).

CHRONOLOGICAL TABLE

1599–1600 Shakespeare's *Julius Cæsar*, *Merry Wives of Windsor*, and *As You Like It* were produced.

1600 Giordano Bruno, Italian scientist, burned at the stake.

1600 William Gilbert published *De Magnete Magneticisque Corporibus et de Magno Magnete Tellure Physiologia Nova.*

1600 *Twelfth Night* and *Hamlet* produced.

1601 The Accademia dei Lincei opened at Rome.

1601 Tycho Brahe, Danish astronomer, died. Athanasius Kircher, German inventor of the magic lantern, born.

1601 Philemon Holland published the first edition of his translation of Pliny's *Natural History.*

1601–2 Shakespeare's *Troilus and Cressida* and Ben Jonson's *Every Man in His Humour* produced.

1602 Otto von Guericke, German scientist, born.

1602–3 *All's Well That Ends Well* produced.

1603 Elizabeth I died. James I succeeded her.

1603 William Gilbert died. Sir Kenelm Digby born.

1603 Geronimo Fabrizio d'Aquapendente published *De Venarum Ostiolis*, describing his discovery of valves in veins of animals.

1604 Sir George Ent, English physician, and Johann Rudolf Glauber, German chemist, born.

1605 Gunpowder Plot. *Don Quixote*, Part I, published.

1605 Francis Bacon published *The Advancement of Learning.*

1605–6 *Macbeth* and *King Lear* produced.

1606–7 *Antony and Cleopatra* produced.

1607–8 *Coriolanus* and *Timon of Athens* produced. Edward Topsell published *A Historie of Four Footed Beasts* (1607) and *The History of Serpents Or the Second Book of Living Creatures* (1608).

1608 John Milton born.

c. 1608 Hans or Franz Lippershey, a Dutchman, made a refracting telescope.

1608–9 *Pericles* produced. Table forks first came into general use (in Italy).

1609–10 Galileo Galilei constructed a telescope and observed several moons of Jupiter.

1609 Charles Butler published *The Feminine Monarchie*, about bees.

1609 John Kepler discovered his first two laws of motion of the planets.
1609–10 *Cymbeline* produced.
1610 Galileo Galilei made a compound microscope.
1610 Henri IV of France (Henry of Navarre) stabbed by Ravaillac.
1611 Marco Antonio de Dominis published *De radiis visus et lucis in vitris perspectivis et iride*, explaining his theory of the rainbow.
1611 The Authorized Version of the Bible published.
1611–12 *The Tempest* produced.
1612–13 Shakespeare's *Henry VIII* produced.
1614 John Napier invented logarithms.
1615 Giambattista della Porta, Italian scientist, died.
1615 *Don Quixote*, Part II, published.
1616 Shakespeare died on April 23, Cervantes on April 18.
1616 John Wallis born.
1618 Sir Walter Raleigh beheaded.
1618 The Fenstersturz (Throwing out of the Window) at Prague.
1618 The Thirty Years War began.
1618 Francesco Grimaldi, discoverer of diffraction of light, born.
1618 John Kepler discovered the third law of motion of the planets.
c. 1619 William Harvey discovered the circulation of the blood in man and animals.
1620 Jean Picard, French scientist, born.
1620 Robert Morison, British botanist, born.
c. 1620 Edmé Mariotte, French scientist, born.
1620 Cornelius van Drebbel, a Dutchman, made a thermometer.
1620 Simon Stevin, Flemish scientist, died.
1620 Francis Bacon published *Novum Organum.*
1621 Francis Bacon, Baron Verulam, Viscount St Albans, and Lord Chancellor, found guilty of taking bribes. Fined £40,000.
1621 Thomas Willis, British scientist, born.
1621 Willebrord Snell van Royen discovered the law of refraction of light.
1622 Molière (Jean-Baptiste Poquelin) born.
1622 The Societas Ereunetica founded at Rostock.
1622 Gasparo Asellio discovered the lacteal glands.
1623 Edmund Gunter published *The Description and Use of the Sector, the Crosse-Staffe and Other Instruments.*
1623 The Dutch massacred some English colonists at Amboyna, East Indies.
1623 Blaise Pascal, French scientist, born.

1623 The First Folio Edition of Shakespeare's plays published.
1624 Thomas Sydenham, British physician, born.
c. 1624 Van Helmont, Flemish chemist, invented the word 'gas.'
1625 Erasmus Bartholinus, Danish scientist, born.
1625 James I died. Charles I succeeded him.
1625 Bacon's *Essays* published.
1626 Francis Bacon, Willebrord Snell van Royen, and Edmund Gunter died.
1627 Robert Boyle, John Ray, and Jacques Benigne Bossuet born.
1627 Shah Jehan became Great Mogul.
1628 William Harvey published *Exercitatio Anatomica de Motu Cordis et Sanguinis in Animalibus.*
1628 Marcello Malpighi, Italian scientist, born.
1628 The Petition of Right presented. The Duke of Buckingham stabbed by Felton.
1628 The House of Commons presented the Petition of Right.
1629 Christian Huyghens, Dutch scientist, born.
1630 The Accademia dei Lincei at Rome closed.
1630 John Kepler died.
1631 Pierre Gassendi, French astronomer, observed a transit of Mercury.
1631 Gustavus Adolphus (Gustavus II) of Sweden defeated Count Tilly at the battle of Breitenfeld.
1632 Christopher Wren born. Gustavus Adolphus killed at the battle of Lutzen.
1632 Anton van Leeuwenhoek, Dutch scientist, born.
1632 Galileo Galilei published *Dialogo sopra i Due Massimi Sistemi del Mundo, Tolemaico e Copernicano.*
1633 Galileo Galilei compelled to abjure Copernician theories. (And yet it does move!)
1634 Philemon Holland published his translation of Pliny's *Natural History* (second edition).
1634 Thomas Moufet's *Insectorum sive minimorum animalium Theatrum* published.
1634 John Milton published *Comus.*
1635 Robert Hooke and Francis Willoughby, British scientists, born.
c. 1635 Thomas Burnet, geologist, born.
1635 Lope de Vega, Spanish dramatist, died.
1636 Pierre Corneille's play *Le Cid* produced.
1637 Ben Jonson died. John Milton published *Lycidas.*
1638 Martin Lister, British scientist, born.

1638 René Descartes, possibly lying in bed, invented analytical geometry.

c. 1638 The first printing press in the United States of America established at Cambridge, near Boston.

1639 Jean Racine born.

1639 Sunday, November 24, the Reverend Jeremiah Horrocks, after preaching his usual afternoon sermon, observed a transit of the planet Venus.

1640 Peter Paul Rubens died. The Long Parliament met. William Wycherley born about this time.

1641 Nehemiah Grew, British botanist, born. Antony van Dyck, painter, died.

1641 Thomas Wentworth, Earl of Strafford, beheaded.

1641 Many English massacred in Ulster.

1641 René Descartes published *Meditationes de Prima Philosophia.*

1642 Abel Tasman discovered Tasmania and New Zealand.

1642 Cardinal Richelieu died.

1642 The English Civil War, Cavaliers versus Roundheads, began. Battle of Edgehill, at which Dr William Harvey looked after the royal children.

1642 Galileo Galilei died. Isaac Newton born on Christmas Day.

1642 English theatres all closed till 1660.

1643–44 Louis XIII of France died. Louis XIV succeeded him.

1643–44 Evangelista Torricelli, Galileo's pupil, invented the mercury barometer.

1644 Olaus Römer, Danish scientist, born. Battle of Marston Moor.

1645 First meetings of virtuosi (the "Invisible College") leading to the formation of the Royal Society.

1645 Battle of Naseby. John Mayow born.

1646 John Flamsteed, English astronomer, and Gottfried Leibnitz, German philosopher, born.

1646 First Civil War ceased.

1647 Evangelista Torricelli died. Denis Papin, French scientist, born.

1648 End of the Thirty Years War. Peace of Westphalia.

1648 Périer, at Blaise Pascal's suggestion, tested a mercury barometer on the Puy de Dôme, September 19.

1648 Marin Mersenne, French scientist, died.

1648 Second Civil War in England.

1649 Charles I beheaded. Oliver Cromwell conquered Ireland.

1649 Richard Baxter published *The Saints' Everlasting Rest.*

1650 Otto von Guericke invented an air pump.

c. 1650 Thomas Savery, British inventor of a steam engine, born.
1650 René Descartes died. Battle of Dunbar.
1651 Thomas Hobbes published *Leviathan*. Battle of Worcester.
1651 William Harvey published *Exercitatio de Generatione Animalium*.
1652 The Collegium Naturæ Curiosorum founded at Schweinfurt.
1652 The Dutch founded Cape Town. Inigo Jones, architect, died.
1653 Izaac Walton published *The Compleat Angler*.
1653 Mlle. de Scudéry published *Artamène ou Le Grand Cyrus*. Barebone's Parliament met.
1654 Pierre Varignon, French scientist, born.
c. 1654 Comenius wrote *Orbis Pictus*, the first picture book for children. It was published in 1658.
1655 Christian Huyghens invented a pendulum clock.
1655 John Wallis published *Arithmetica Infinitorum*.
1655 Jamaica occupied by the British.
1655 Christian Huyghens discovered the rings of Saturn.
1656 Edmund Halley was born.
1656 John Tradescant published *Museum Tradescantium*.
1656 James Harrington published *The Commonwealth of Oceana*.
1656–57 Blaise Pascal published *Les Lettres Provinciales*.
1657 The Accademia del Cimento was opened at Florence.
1657 *Experimentia Nova (ut vocantur) Magdeburgica de Novo Spatio*, describing von Guericke's work on atmospheric pressure and vacua, published by G. Schott.
1657 William Harvey died.
1657 Fountain pens manufactured in Paris.
1658 Thomas Hobbes published *Elementorum Philosophiae* (*Sectio II de Homine*).
1658 Oliver Cromwell died.
1659 Molière produced *Les Précieuses Ridicules*.
1659 Robert Boyle published *New Experiments Physico-Mechanical touching the Spring of the Air* (containing Boyle's Law).
1659 Aurungzeb became the Great Mogul (after imprisoning his father).
1660 The Restoration. The King, Charles II, enjoys his own again.
1660 Athanasius Kircher invented a magic lantern.
1661 Marcello Malpighi used a microscope to study cell structure.
1661 James Gregory, Scottish scientist, invented a reflecting telescope.
1661 Otto von Guericke invented a manometer or pressure gauge, which he described in *Brief an Pater Schott*.

1661 Thomas Salusbury published Galileo's *Dialogues* in English.
1661 Robert Boyle published the *Sceptical Chymist*.
1661 Robert Hooke published *An Attempt for the Explication of the Phenomena*, etc. Daniel Defoe born.
1662 The Royal Society of London founded by royal charter.
1662 Blaise Pascal and Francesco Grimaldi died. Richard Bentley born.
c. 1663 Samuel Butler wrote *Hudibras*.
1663 The Marquis of Worcester published *Century of the Names and Scantlings*.
1663 James Gregory published *Optica Promota*.
1664 Molière produced *Tartuffe*. Matthew Prior born.
1664 Thomas Willis published *Cerebri Anatome*.
1665 Francesco Grimaldi's book *Physico-Mathesis de Lumine, Coloribus et Iride* published.
1665 Sir Kenelm Digby died.
1665 John Woodward, geologist, born.
1665 The Duke of La Rochefoucauld published *Réflexions ou Sentences et Maximes Morales*.
1665 Robert Hooke published *Micrographia*. Mr Samuel Pepys bought a copy on January 21, 1665 (new style).
c. 1665 Isaac Newton invented his method of fluxions.
1666 John Dryden published *Annus Mirabilis*.
1666 The Académie Royale des Sciences at Paris founded.
1666 Isaac Newton discovered the law of gravitation, but remained silent. He also carried out his researches on optical dispersion.
1667 William Whiston, scientist and historian, born.
1667 John Milton published *Paradise Lost*. Jonathan Swift born.
1667 The Accademia del Cimento at Florence closed. Its work was published in the same year with the title *Saggi di Naturali Esperienze*.
1667 Thomas Sprat published *A History of the Royal Society*.
1668 John Mayow published *Tractatus duo, de Respiratione et de Rachitide*.
1668 La Fontaine's *Fables* published. Alain René Lesage, author of *Gil Blas*, born. Johann Rudolf Glauber died.
1668 Robert Hooke published *A Discourse on Earthquakes*.
1668 Isaac Newton made a reflecting telescope.
1669 Niels Stensen, alias Steno, a Dane, published a work on crystals and fossils called *De solido intra solidum naturaliter contento dissertatio prodromus*.

1669 Richard Lower published *Tractatus de Corde*.
1669 Rembrandt died.
1670 John Mayow discovered "fire air" (oxygen).
1670 Samuel Morland made a speaking tube. William Congreve born.
1670 The *Pensées* of Blaise Pascal were published.
1671 John Milton published *Paradise Regained* and *Samson Agonistes*.
1672 Otto von Guericke described an electrical machine.
1672 Joseph Addison born.
1672 Isaac Newton's first memoir, on optics, published in the *Philosophical Transactions of the Royal Society*.
1672 Nehemiah Grew published the *Anatomy of Vegetables Begun*.
1672 Francis Willoughby died.
1672 Thomas Willis published *De Anima Brutorum*.
1673 Christian Huyghens published *Horologium Oscillatorium sive de Motu Pendulorum*.
1673 Molière produced *Le Malade Imaginaire*.
1673 Samuel Morland published *The Description and Use of Two Arithmetick Instruments*.
1674 John Mayow published *Tractatus quinque Medico-Physici*.
1674 William Noble and Thomas Pigot studied the vibrations of strings, at Oxford.
1674 John Milton died.
1675 Thomas Willis died. The Duc de Saint-Simon born.
1676 Olaus Römer measured the velocity of light in free space.
1676 *Ornithologiæ Libri Tres*, by John Ray and Francis Willoughby, published.
1676 Robert Hooke published the anagram *ceiiinosssttuu* representing *ut tensio sic vis* (Hooke's law of elasticity).
1677 Francis Glisson died.
1678 Robert Hooke published *Lectures de Potentia Restituva; or of Spring*.
1678 John Bunyan published *Pilgrim's Progress*, Part I.
1678–81 Martin Lister published *Historiae Animalium Angliae tres tractatus*.
1678 Christian Huyghens proposed a wave theory of light.
1679 Edmund Halley published *Catalogus Stellarum Australium*.
1679 Thomas Hobbes and John Mayow died.
1680 Samuel Butler died.
1680 Athanasius Kircher died.
1681 Thomas Burnet published *Telluris Theoria Sacra*.
1681 John Dryden published *Absalom and Achitophel*.

1682 Jean Picard measured the size of the earth with increased precision.
1682 Nehemiah Grew published *The Anatomy of Plants.*
1682 Edmund Halley foretold the return of a comet every seventy-six years.
1683 Robert Morison died.
1684 Edmé Mariotte died.
1684 Richard Waller published *Essayes of Natural Experiments*, a translation of the *Saggi di Naturali Esperienze* (1667). Part II of *Pilgrim's Progress* published.
1685 Charles II died. James II succeeded him.
Monmouth's Rebellion.
1685–92 Martin Lister published *Historia sive Synopsis Methodica Conchylorum.*
1686 Otto von Guericke died.
1686 *De Historia Piscium. Libri quattuor*, by John Ray and Francis Willoughby, published.
1687 Isaac Newton published *Philosophiae Naturalis Principia Mathematica.*
1688 The Glorious Revolution. James II abdicated. Alexander Pope born.
1689 William and Mary succeeded.
1689 Thomas Sydenham and Sir George Ent died. Samuel Richardson born.
1689 Montesquieu (Charles Louis de Secondat, baron de La Brède) born.
1691 Robert Boyle died.
1693 *Synopsis Methodica Animatium Quadrupedum et Serpentine Generis Vulgarum*, by John Ray and Francis Willoughby, published. *Three Physico-Theological Discourses* published by John Ray alone.
1694 Joseph Tournefort described his system of classification of plants.
1694 Voltaire (François Marie Arouet) born.
1695 John Woodward published *Natural History of the Earth and Terrestrial Bodies.*
1695 Christian Huyghens died.
1696 William Whiston published *A New Theory of the Earth.*
1698 Thomas Savery took out a patent for a steam-engine.
1698 Erasmus Bartholinus died.

INDEX